METASTABLE IONS

METASTABLE IONS

R.G. COOKS, J.H. BEYNON, R.M. CAPRIOLI and G.R. LESTER

ELSEVIER SCIENTIFIC PUBLISHING COMPANY

Amsterdam — London — New York

1973

ELSEVIER SCIENTIFIC PUBLISHING COMPANY
335 Jan van Galenstraat
P.O. Box 1270, Amsterdam, The Netherlands

AMERICAN ELSEVIER PUBLISHING COMPANY, INC.
52 Vanderbilt Avenue
New York, New York 10017

Library of Congress Card Number: 72-97419

ISBN 0-444-41119-4

Printed in The Netherlands

Acknowledgements

The permission of the copyright owners and the authors for the use of published diagrams is gratefully acknowledged: Heyden and Son, Ltd. for Fig. 18 (from ref. 282), Fig. 27 (from ref. 49), Fig. 33 (from ref. 54), Fig. 65 (from ref. 31), Fig. 90 (from ref. 168), Fig. 91 (from ref. 105) and Figs. 97 and 98 (from ref. 59); the American Chemical Society for Figs. 52, 74, 80 and 81 (from ref. 12), Fig. 69 (from ref. 89) and Figs. 99 and 100 (from ref. 62); Taylor and Francis, Ltd. for Figs. 20 and 70 (from ref. 110); North-Holland Publishing Company for Fig. 45 (from ref. 119); the American Institute of Physics, Inc. for Figs. 71 and 72 (from ref. 157); the Chemical Society (London) for Fig. 87 (from ref. 57); the Royal Society (London) for Fig. 89 (from J.H. Beynon, R.G. Cooks and R.M. Caprioli, *Proc. Roy. Soc. Ser. A*, 327 (1972) 1); Technical Publishing Company for Fig. 36 (from ref. 56); Redaktion der Zeitschrift für Naturforschung for Figs. 62 and 63 (from ref. 132). Several figures to which Elsevier holds copyright have also been used; these are Fig. 40 (from ref. 58), Figs. 59—61 (from ref. 46), Fig. 55 (from K. Levsen and H.D. Beckey, *Int. J. Mass Spectrom. Ion Phys.*, 7 (1971) 341), Fig. 64 (from ref. 30) and Fig. 92 (from ref. 53).

We record, with thanks, our appreciation to Janet Shaw and Pat Walton for their stenographic assistance, to Larry Eckstein for preparing the figures and to our wives for putting up with us.

Contents

"You, the bold seekers and tempters, and whoever embarks with cunning sails on terrible seas — you, drunk with riddles, glad of the twilight, whose soulflutes lure astray to every whirlpool, because you do not want to grope along a thread with cowardly hand; and where you can *guess*, you hate to *deduce*."

Nietzche, *Thus Spake Zarathustra*

Introduction

"I will sing of facts; but some will say that I invented them."

Ovid, *Fasti*

The purpose of this book is to bring together in a single small volume, the information which is scattered throughout the literature concerning the fragmentation of metastable polyatomic positively charged ions. Such information has been acquired largely through the application of mass spectrometry to problems in organic chemistry. Although an increasing number of mass spectrometers is being used for research in physical chemistry, by far the largest number of instruments is in organic chemistry laboratories where they are used mainly to help in elucidating the structures of complex organic substances. Several recent books [33,42,66,74,138,187,226,244,248] have been concerned with the interpretative procedures used to deduce structural formulae from mass spectra, but they deal only incidentally with metastable ions. The authors believe there is so much useful and detailed information to be obtained from a careful study of metastable ions in mass spectra that it is worthwhile to attempt to review the present level of knowledge. They further believe that an increased awareness among researchers of the extra information available through study of the reactions of metastable ions will lead to an extension of the range of complex organic compounds that can be characterized by mass spectrometry, that it will stimulate new applications of mass spectrometry to chemistry and that it may lead to the development of new instruments better suited to presenting all the information hidden within the mass spectrum.

The book does not set out to be a fully comprehensive treatise. Rather the aims have been to present a simple illustrative account of the applications already envisaged, to explain in detail how the reactions of metastable ions can be observed and to give an introductory account of the underlying theory both for polyatomic organic and for simple inorganic ions. A description of some of the instrumentation commonly used to study metastable ions has been included together with some important properties of magnetic and electric sectors. Emphasis has been placed on the measurement of ion kinetic energies and on the significance of changes that can be detected in

this energy as various ionic reactions proceed. Although metastable
ions are ionic species that decompose spontaneously and unimolecularly, a description of selected reactions that are induced by collision of ions and neutral gas-phase molecules has also been included.

The study of the reactions of metastable ions is still in an early
stage of development, but already some jargon has become widely
accepted. In particular, the signal recorded when the daughter ions
formed from the fragmentation of metastable parent ions is being
plotted has become known as a metastable peak. We have used this
term as a convenient shorthand in the belief that its meaning is
universally understood. We have also used a conveniently short expression when discussing the amount of internal energy that is converted into translational energy during the fragmentation of a metastable ion. The shape of the metastable peak observed in such a case
reflects this energy transfer but the detailed relationship between
peak shape and the distribution of kinetic energy releases is not fully
understood. Therefore, whenever an estimate of this energy is given,
it should be understood that the calculation has been made relative
to the width of the metastable peak at half-height unless another
method is expressly referred to. We have tried to avoid assigning
detailed structures to complex ions without supporting evidence but
in several cases a structural formula provides a convenient notation
for expressing an idea or discussing a reaction and, indeed, it sometimes seems that it provides the only effective means of translating
the results of a study of the mechanism of a particular reaction.

Because the book is selective in reporting the advances that have
been made since the inception of the subject, there is a tendency to
quote mainly from recent work and to under-emphasize the important part played by a small number of researchers in laying the
foundations of the subject since the observation of "knots" in his
parabolic curves by Thomson. Particular mention must be made of
the contribution of Nier to the instrumentation that is so widely
used, of Hipple and Condon who recognized metastable peaks in
1945 and carried out the first quantitative measurements on them,
and of a small number of physical/organic chemists, among whom
Jennings and McLafferty have been especially prominent, who by
their keen observation and lively imagination have stimulated the
development of the subject over a broad front. The contribution of
many others can be acknowledged only in the references.

It is hoped that the book will be of use as a textbook for the
undergraduate or postgraduate student beginning research in mass
spectrometry and that it will also provide something of interest for

the more advanced researcher and will help to stimulate his thinking along some new directions.

Chapter 1

Instrumentation

*"Shakespeare says, we are creatures that look before and after:
the more surprising that we do not look round a little, and see
what is passing under our very eyes."*

Carlyle, *Sartor Resartus*

In a mass spectrometer, the sample to be analyzed is generally
introduced into an ionization chamber in which ions characteristic of
the sample molecules are produced. By the action of electric (and,
usually, magnetic) fields, these ions are sorted according to their
mass-to-charge ratios and a plot of the relative abundances of the
ions against these ratios constitutes the mass spectrum.

A great variety of types of mass spectrometer has been described,
many built for special applications, and the reader should refer to
other books [1,33,67,139,151,161,183,229,272] for a description of these. In organic mass spectrometry, most of the work has
been carried out using instruments employing sector-shaped magnets
and electron bombardment ion sources and a fairly complete account
of such instruments will be given here. This discussion is essential to the
understanding of the formation and reactions of metastable ions. In
describing experimental results reported in the literature, reference
will, however, be made from time to time to other kinds of mass
spectrometers.

THE IONIZATION CHAMBER

A typical arrangement of an electron bombardment ion source is
shown in Fig. 1. This type of ion source was first introduced by
Dempster and described by him in 1921 but it has since been improved in the later designs of many other workers, notably by Nier
[219]. Electrons are produced from an incandescent helical filament or
ribbon usually made from tungsten or rhenium. The electrons are
accelerated towards the ionization chamber and enter it through a
system of collimating slits (not shown in the figure). The kinetic
energy of the electrons can be varied simply by changing the potential difference between the filament and the ionization chamber. The

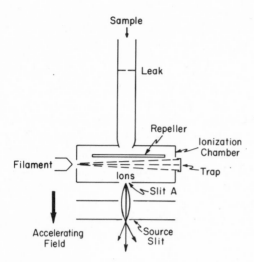

Fig. 1. Electron impact ion source.

region within the ionization chamber is essentially free of electric fields and the electrons traverse it at constant velocity. Electrons which pass through a slit at the far end of the ionization chamber are accelerated towards an electrode at a potential higher than that of the ionization chamber and collected. This electrode is known as the "trap" or target and either the current to this trap or the sum of this current and the current flowing to the ionization chamber walls (the total emission current) can be used to regulate the filament current and so maintain a constant rate of ion production. In order to constrain the electron current passing through the ionization chamber into a narrow beam, it is usual (but not essential) to provide a magnetic field from a permanent magnet along the direction of the electron flow. Electrons with a component of velocity at right angles to the magnetic flux move in tight helices along the axis of the field. For a magnetic flux density of 3×10^{-2} T (300 G), electrons of 50 eV energy move on a helical radius of 0.34 mm with a pitch of 7 mm. Commonly, the length of the ionization chamber is about 20 mm.

To raise a typical filament to its emission temperature requires a current of the order of 3 A and the potential drop along the length of the filament is of the order of 1 V. Mainly for this reason, the electron beam is not mono-energetic but has a small spread of energies about the mean value. The emission temperature of the filament is greater than $2000°C$ and the wattage necessary to maintain it at this temperature raises the temperature of the entire ionization chamber to

100—150°C. This is sufficient to cause decomposition of some organic compounds and may require the use of a cooled source or an alternative source of ionization, such as energetic photons, in special circumstances. For this same reason, sample which has come into contact with the filament surface must not be allowed to re-enter the ionization chamber and this requires that the slits in the ionization chamber be small and the pumping rate outside the chamber high. Sample is usually introduced in the form of a vapor along the tube shown in Fig. 1 *via* a pin-hole leak. A variety of sample-handling systems has been described in which a weighed amount of sample can be vaporized into a reservoir and brought to a suitable pressure for passing through the leak. Alternatively, samples which are less volatile can be evaporated at an appropriate rate from a small "probe" inserted through the wall of the ionization chamber (not shown in the figure). It is not possible to measure directly the sample pressure within the ionization chamber. Pressure measurements are usually made by means of an ionization gauge located as near as possible to the outside of the chamber. The ratio of the pressure within the chamber to the recorded pressure will depend upon the design of the ionization chamber and the pumping rate outside it, but commonly the ratio will be of the order of 10:1. A typical sample pressure within the ionization chamber is 1.3×10^{-3} Pa (10^{-5} torr). The mean free path at this pressure will be of the order of 10^5 mm so that collisions between gas molecules as they traverse the ionization chamber are very unlikely. The collision cross-section between ions and molecules will usually be several orders of magnitude greater than between molecules but, nevertheless, the probability of collision over a single traverse of the ionization chamber will be very small. Molecules will, however, collide with the walls and when they do they will be re-emitted from the wall at an angle dependent on the surface conformation and independent of their angle of incidence. The probability of a molecule escaping from the ionization chamber and being pumped away will depend upon the relative areas of the slits and wall surfaces and in a typical ionization chamber a molecule will collide with the walls about 50 times before escaping. For this reason, it is important that the temperature of the walls be kept low.

One or two repeller electrodes are usually located within the ionization chamber. When a small positive voltage is applied to them, any ions formed by the electrons will be repelled towards the slit A. Ions which do not escape through this slit will, of course, lose their charge on the walls, but the probability of escape is much greater for an ion than for a neutral molecule which cannot be influenced by the field of the repeller electrodes. The theory of extraction of ions is beyond the scope of this book, but it should be noted that the efficiency of extraction is known to fall off as the source pressure is increased.

A typical filament operates at a total emission current of about 10^{-4} A and about 25% of these electrons will pass through the ionization chamber and be collected on the trap. One electron/sec corresponds to a current of 6×10^{-20} A, so about 4×10^{14} electrons/ sec are passing through the ionization chamber. The probability that one of these electrons will make an ionizing collision with a gas molecule [of, say, molecular weight 100 at a pressure of 1.3×10^{-3} Pa (10^{-5} torr)] is only of the order of 2 in 10^6. About 10^{15} gas molecules/sec will be passing through the ionization chamber at this pressure, so that of the order of 2×10^9 ions/sec will be formed, corresponding to a total ion current of about 10^{-10} A. It will be important to bear these figures in mind in considering the reactions of metastable ions contained in this beam.

Let us now consider the time scale of the major ionization event occurring within the ionization chamber when electrons of 50 eV energy are employed. If one of these electrons can pass sufficiently close to one of the orbiting electrons in a sample molecule, it will repel this electron out of the molecule to give a positively charged molecular ion. The reaction can be written

$$M + e \rightarrow M^{+} \cdot + 2\,e \tag{1}$$

A 50 eV electron travels with a velocity of 4.2×10^8 cm/sec and it will thus traverse a molecular diameter of the order of 10^{-7} cm in a time of 2.4×10^{-16} sec. We can expect the act of ionization to be completed in this order of time. The fastest molecular vibration, a C—H stretching vibration, has a period of the order of 10^{-14} sec and so we may consider all the atoms to be effectively at rest during ionization; thus, we are dealing with a "vertical" rather than an "adiabatic" ionization process. The ion formed will often be in an electronically excited state and in Appendix I we shall consider how this excess energy can redistribute itself into vibrational and other forms. Due to its positive charge, the ion will now come under the action of the repeller plates and move out of the ionization chamber. Assuming that it starts from rest, is of mass 100 and travels 0.1 cm before leaving the ionization chamber, the time involved will be 1.4×10^{-6} sec if it acquires an energy of 1.0 eV from the repeller plates and penetration of the accelerating field.

THE ION ACCELERATOR

The ions which leave the ionization chamber under the influence of the repeller electrodes do so with a very small kinetic energy corresponding to about one electron volt. Outside the ionization chamber, however, they come under the action of a strong electric field corresponding to a potential drop of several thousand volts over a distance of the order of 1 cm and are accelerated towards a plate containing a slit, known as the source slit, entrance slit or beam-defining slit. The ions usually studied by mass spectrometry carry a positive charge and, in order to keep the source slit at ground potential, the ionization chamber is operated at a positive potential above ground of several thousand volts. The ion accelerator usually contains some focusing plates designed to concentrate the ion beam on the defining slit and is arranged in such a way that penetration of the accelerating field into the ionization chamber is minimized. If the fields within the ionization chamber can be kept small, the ions formed in the narrow electron beam will be produced at points of the same electrical potential and, after acceleration, will all possess essentially the same kinetic energy. In order to keep the *fractional* spread in energy of the ion beam as small as possible, high accelerating voltages are used, approximately 10 kV being used in high-performance instruments.

The kinetic energy of an ion of mass m and charge e accelerated through a potential drop V is given by $mv^2/2$ where v is the terminal velocity and

$$\tfrac{1}{2} mv^2 = eV \tag{2}$$

A singly charged ion of mass 100 falling through an accelerating voltage of 10 kV will acquire a velocity of 1.4×10^7 cm/sec. If the accelerating region is 2 cm long, it will pass through the analyzer entrance slit 3×10^{-7} sec after leaving the ionization chamber. The beam of ions that has a kinetic energy equal to the full accelerating energy is known as the "main ion beam".

THE MAGNETIC ANALYZER

In order to analyze the ion beam issuing from the accelerating region, it is necessary to separate out the components of different mass and to focus each of these discrete mass beams so that their intensities can be measured. This is usually accomplished by the use of a magnetic field in a direction perpendicular to the direction of ion motion. An ion of mass m, charge e and velocity v injected into a magnetic flux B in this way will follow a circular path of radius r given by

$$r = \frac{mv}{Be} \qquad\qquad\qquad (3)$$

The radius can be seen to be proportional to the *momentum* of the ions.

An ion beam of uniform kinetic energy and containing only ions of mass m which diverges from a slit with a half-angle α as shown in Fig. 2 will come to perfect focus at the slit as the ions pass through the point of origin on each revolution. They will also come to an

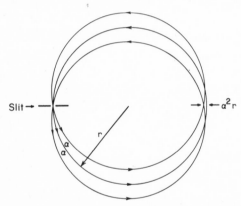

Fig. 2. Focusing action of a magnetic field on a beam of ions all of mass-to-charge ratio m/e and of uniform kinetic energy. The beam diverges from a slit with a half-angle α.

approximate focus after a half-revolution in the magnetic field, the width of the beam at this position being equal to $\alpha^2 r$. In the absence of focusing, the width of the beam would be proportional to α; the condition in which the width of the beam is proportional to α^2 is called "first-order" focusing, when the beam width is proportional to α^3 we have "second-order" focusing and so on. Magnetic analyzers of $180°$ angle have been used in several commercial mass spectrometers and Fig. 3 shows how such a field will separate an ion beam of uniform kinetic energy into its constituent masses. From eqn. (2) we have that

$$v = \left(\frac{2\,eV}{m}\right)^{1/2}$$

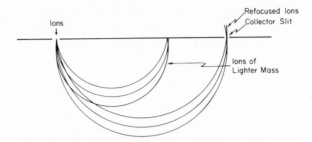

Fig. 3. Focusing action of a $180°$ magnetic analyzer on a mono-energetic beam containing ions of two different mass-to-charge ratios. Note that the ions do not come to a perfect focus.

and thus

$$mv = (2eVm)^{1/2} \tag{4}$$

The momentum, and thus the radius of curvature in the magnetic field, will vary as the square root of the mass. By changing the magnetic field, the ions of different mass can be brought successively to a focus at the recording slit.

Because the magnetic field focuses a mono-energetic beam of ions diverging from the source slit, it is said to give a "direction focus". An ion beam of different energy will follow a different radius of curvature in the magnetic field and be brought to a different focus. For this reason, the magnetic field is said not to give a "velocity focus" but to be "single-focusing" for direction only.

Equations (3) and (4) can be combined to give the relationship

$$\frac{m}{e} = \frac{B^2 r^2}{2V} \tag{5}$$

and this can also be written in the form

$$2\frac{\delta r}{r} = \frac{\delta m}{m} - \frac{2\delta B}{B} + \frac{\delta V}{V} \tag{6}$$

which shows the effect of small changes in the various parameters on the radius of curvature of the ion path.

The width of the image of a line source is given by $d = r(\delta m/m)$ when B and V are constant and from this the mass dispersion can be determined. This can be defined as the value of d when $\delta m/m = 0.01$ and is given by

$$D_m = \frac{r}{100} \tag{7}$$

The resolving power R_m of the magnet is a measure of its ability to separate adjacent masses m and $(m + \delta m)$. Ignoring all distortions of the image, the geometrical limit of resolving power is given for symmetrical object and image distances by

$$R_m = \frac{m}{\delta m} = \frac{r}{S_1 + S_2} \tag{8}$$

where S_1 and S_2 are the widths of the object and image slits. Distortions in the image will add a further term to the denominator of this expression. For a value of r of 30 cm the effective limit of resolving power is of the order of 5000.

It was first shown in 1933 by Barber [17] and later by Stephens [250, 251] that this lens action of a 180° magnetic field is a special case of the focusing action of a wedge-shaped field. Fig. 4 illustrates the focusing action of a wedge-shaped or sector field of 90° angle. It

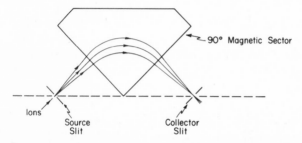

Fig. 4. Focusing action of a wedge-shaped (sector) magnetic field on a mono-energetic beam of ions, all of the same mass-to-charge ratio.

can be shown that if the ion beam enters and leaves the magnetic field in a direction perpendicular to the field boundaries, then the point of origin of the ions, the apex of the field and the point of focus are collinear. It can also be shown that the dispersion obtained is independent of the sector angle, but the path length traveled by the ion beam will increase as the sector angle is reduced. The sector angle chosen will depend upon the ion path length required for the particular applications envisaged for the instrument and may also be influenced by considerations such as the increased weight and cost of magnets having large sector angles. Most instruments built to date have used angles of 60° or 90°.

A magnetic sector produces a focusing action only in a direction perpendicular to the magnetic field (commonly known as the y direction). It thus behaves as a cylindrical lens. With the simple system described there is no focusing action in the direction parallel to the

magnetic field (the z direction). The x, y and z directions are illustrated in Fig. 5. Various arrangements have been described to give a
measure of z-axis focusing and also to produce better than first-order
y-axis focusing [134, 178] but these are beyond the scope of the
present book.

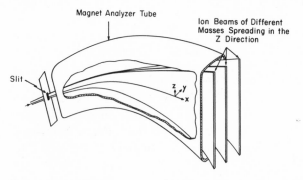

Fig. 5. Spreading of an ion beam in a direction parallel to that of the magnetic
field (z direction). The ion beam travels along the x direction and the focusing
action of this field (and also of the electric sector field) operates only in the y
direction.

THE ENERGY ANALYZER

   In order to reduce the image spread in the focused beam issuing
from the magnetic sector, early mass spectrometers employed some
form of "velocity selector" to ensure that the beam entering the
magnetic field was as homogeneous in energy as possible. These velocity selectors produced no focusing action and so they reduced
considerably the intensity of the transmitted beam. Focusing energy
analyzers can also be employed and they will be considered further
because they are used in most high-performance instruments as will
be seen below. Let us consider first an electric sector such as is
shown in Fig. 6 which produces a radial electric field, that is to say, a
field in which the potential $V(r)$ and the radius vector r are related
by the expression

$$r \cdot \frac{\partial V(r)}{\partial r} = \text{constant} \tag{9}$$

If the outer plate is made positive with respect to the inner plate, and
a beam of ions of various energies is injected midway between the
plates and perpendicular to the direction of the electric field, there
will be some ions which will describe a circular trajectory along the

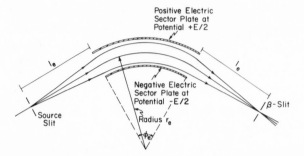

Fig. 6. Direction focusing action of a radial electric field of angle ϕ_e. All ions of the appropriate kinetic energy (proportional to E) are brought to a focus at the β-slit whatever their masses.

curve of radius r_e which is a line of equipotential. This condition is met when the kinetic energy of these ions, $mv^2/2$, is such that the electrostatic force on the ions is exactly balanced by the centrifugal force. Thus we can write

$$\frac{mv^2}{r_e} = xeF \tag{10}$$

where F is the field strength and x the number of charges carried by the ions. From eqn. (2), we have that

$$mv^2 = 2xeV$$

and thus

$$r_e = \frac{2V}{F} \tag{11}$$

The mass m and the number of charges x carried by the ions do not appear in this equation so that ions of all masses carrying any number of charges will follow the same radius through the electric sector, provided only that they have been accelerated through a voltage V. Only ions with the predetermined energies xeV will describe the radius r_e. Ions of greater or smaller energies will describe greater or smaller radii so that the radial electrostatic field behaves analogously to a prism in producing an energy dispersion in the ion beam and can be used as an energy filter if a slit is placed behind the sector.

A radial electric field will also produce a focusing action on a diverging mono-energetic beam of ions. The derivation of the equations connecting object and image distances with the sector angle and the radius r_e is beyond the scope of this book and the reader is referred to the specialized treatments quoted for further details. For equal object and image distances l_e and a sector angle ϕ_e, the relationship

$$l_e = \frac{r_e \left[1 + \cos\left(2^{1/2}\,\phi_e\right)\right]}{2^{1/2}\sin\left(2^{1/2}\,\phi_e\right)} \tag{12}$$

holds. For $\phi_e = 60°$, $90°$ and $127.17°$ the values of l_e are $0.65r_e$, $0.35r_e$ and zero, respectively. This is illustrated in Fig. 7. Radial electrostatic fields behave only as cylindrical lenses and produce no

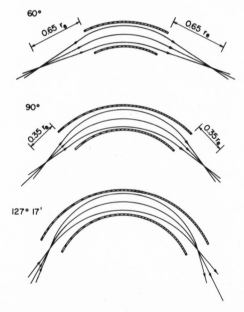

Fig. 7. Illustration of the different focal lengths of radial electric fields all of radius, r_e, but of different angles, ϕ_e.

focusing action in the z direction (perpendicular to the plane of deflection of the ions). "Spherical" condenser lenses can be used in special cases to produce z-axis focusing. Alternatively, plates at a positive potential can be placed in planes parallel to the deflection plane, above and below the plates giving rise to the radial electric field, so as to limit spreading of the beam in the z direction as it passes through the electric sector. From eqn. (11), one obtains

$$\frac{\delta r_e}{r_e} = \frac{\delta V}{V} - \frac{\delta F}{F} \tag{13}$$

The energy dispersion can be obtained by considering the lateral displacement, D_e, in the image plane, of ions entering the field along the central ray but with an energy corresponding to an accelerating voltage $(V + \delta V)$. For symmetrical object and image distances, this is given by

$$D_e = \frac{r_e \delta V}{V} \tag{14}$$

The energy resolution, also for the symmetrical case, is given by the relationship

$$R_e = \frac{r_e}{S_1 + S_2} \tag{15}$$

where S_1 and S_2 are the widths of the object and image slits and all distortions in the image are neglected. In the practical case, such distortions will add a further term to the denominator in eqn. (15). For a value of $r_e = 30$ cm, a practical limit of energy resolution is about 5000.

DOUBLE-FOCUSING INSTRUMENTS

Early mass spectrometers were either direction focusing or velocity focusing. In order to obtain the best performance possible, some special provision was usually made to compensate for the fact that one form of focusing was missing. For example, in the magnetic sector mass spectrometers which give direction focusing only, an electron bombardment ion source was used which produced only a very small spread of ion energies. The idea of a double-focusing instrument which would incorporate both properties was suggested in 1929, but the construction of high-performance instruments awaited better understanding of the ion optics involved. What was needed was an arrangement in which electric and magnetic sectors were combined in such a way that the energy dispersion produced by one sector was counterbalanced by that of the other. At the same time, directional focusing was to be maintained throughout the entire path. Several instruments giving first-order double focusing were under construction when, in 1934, Mattauch and Herzog [178] published a complete theory of achieving double focusing using a radial electric field and a uniform magnetic field with straight boundaries. A feature of this very important work was that it was shown that the focusing conditions could be derived in a familiar geometrical optics form.

Two designs of double-focusing instruments employing sector fields have become widely used and both are available as commercial instruments. The first of these is due to Nier and his colleagues and is shown in Fig. 8. This design gives second-order direction focusing coupled with first-order velocity focusing. Using an ion radius of 30 cm in the magnet, a resolving power greater than 10^5 has been achieved and mass measuring accuracy of 1 in 10^8 can be reached over a small mass range.

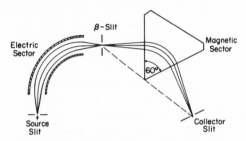

Fig. 8. Arrangement of the electric and magnetic fields for a double-focusing mass spectrometer of Nier—Johnson type geometry.

The other widely used commercial design is named after Mattauch and Herzog and is shown in Fig. 9. In this instrument deflection in the magnetic field is in the opposite direction to that in the electric field and there is no intermediate focal point between the fields. A particular feature of this design is that it focuses all masses in a plane in which a photographic plate can be placed. The mass spectrum is produced as a series of lines, the position of any line being dependent on the mass-to-charge ratio of the ions which form it, the density of the line depending on the intensity of the ion beam. A resolving power of the order of 10^5 can again be obtained with comparable mass-measuring accuracy to that described above.

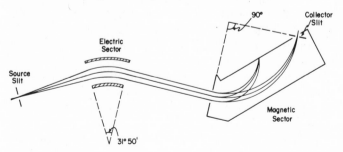

Fig. 9. Arrangement of the electric and magnetic fields for a double-focusing mass spectrometer of Mattauch—Herzog geometry.

Alternatively, conventional recording, peak by peak, using a collector slit and a multiplier is possible with either design, the spectrum being scanned by varying the magnetic field.

MODIFIED INSTRUMENTATION FOR STUDIES OF METASTABLE IONS

The design parameters for an instrument to be used largely or entirely for the study of metastable ions differ from those for equipment to be used, for example, for high-resolution mass spectrometry. High sensitivity for the study of the weak beams of product ions becomes of paramount importance and this requirement is reflected in the desirability of having an abundant source of ions, large slits and very sensitive detectors capable of following the arrival even of single ions. It will be shown in Chapter 3 that there are advantages in arranging the electric and magnetic sectors so that the beam passes through the magnetic field first; each mass component of the beam can then be studied separately. In order to study very small energy releases in the decompositions of metastable ions, double focusing is necessary and the energy resolution of the electric sector needs to be as high as possible. This requires the use of narrow slits and there may be a further limitation on slit length coupled with the necessity to collimate the beam of ions entering the electric sector if it is required to study energy releases falling into a narrow range. These latter features are entirely different from those desirable in high-sensitivity studies. Thus, as for any instrument, the design criteria will be dependent upon the direction taken by the research. The basic features of the equipment necessary for the various studies can only be properly appreciated when the experiments to be conducted and the magnitudes of the various effects that can be produced are themselves understood in detail. For this reason, the detailed design of special instrumentation, including that in the authors' laboratory, will be discussed later in the book when the fundamental properties of the ions have been derived. Such features as the number of field-free regions, the reaction time interval sampled, the value of differential pumping, the scanning of the high voltage and electric sector voltage in various ways, the use of charge-exchange methods to isolate particular groups of ions and to study the energetics of their reactions, details of peak shape and position, will all be developed later. The field of study is still relatively new and has become truly quantitative only in the past few years; as the field develops, the instrumentation is developing alongside it.

Chapter 2

Types of ions formed in a mass spectrometer

*"All things must change
To something new, to something strange"*

Longfellow, *Kéramos*

Many different kinds of ions occur in the mass spectra of organic compounds. The main terms used to describe them and the various ways in which they can be formed are listed below.

POSITIVELY CHARGED IONS

A positive ion is a molecule or free radical from which one or more electrons have been removed so that it carries a net positive charge. Study of these ions forms the basis of mass spectrometry. The commonly used methods of ionization generally form positive ions in great variety and abundance and most of the ionic species discussed below fall into this general category.

NEGATIVELY CHARGED IONS

Under the conditions usually used in mass spectrometers, negative ions are also formed, although in low abundance. A negative ion is a molecule or free radical to which has been added one or more electrons. In the process in which a positive ion is formed from a neutral species, an electron has to be expelled. This reaction can be brought about by the use of bombarding electrons (or other species) having energies ranging from the order of 10 eV to nuclear energies. In contrast, the process of electron attachment requires a much more carefully controlled interaction between the electron and the neutral. The enthalpy of the negative ion formed will include a term that describes the stability in terms of the binding energy of the extra electron. This is known as the electron affinity of the neutral molecule or free radical. If the neutral is bombarded with mono-energetic electrons of gradually increasing energy, a sharp peak may be observed between an electron energy of zero and a few eV. The position of this peak is determined by the electron affinity; its width of only a few meV reflects the fact that electron attachment is a resonance process.

The negative ions that are observed in low abundance (about 0.1—1.0% of the abundance of positive ions) in mass spectra produced by 70 eV bombarding electrons arise by secondary processes, mainly from very low energy photoelectrons that exist inside the source.

Negative ions can also be formed by the process of ion pair production and this is mentioned again later in this chapter.

MOLECULAR IONS*

Unless otherwise stated, the term molecular ion designates a positively charged ion; that is to say, an ion formed from a molecule by removal of one electron. The process for formation of a singly charged molecular ion by electron bombardment can be written

$$M + e \rightarrow M^{+\cdot} + 2\,e$$

where M represents the molecule and $M^{+\cdot}$ the molecular ion. Most organic molecules contain an even number of electrons, so that when a single electron is removed an odd-electron ion results. This is indicated by the dot placed alongside the symbol for the positive charge [74, 241]. If, however, the bombarded molecules are free radicals, as in the case of nitric oxide, the ionization process would be represented by

$$NO^{\cdot} + e \rightarrow NO^{+} + 2\,e$$

and gives rise to an even-electron ion. The process of ionization by electrons has been mentioned in Chapter 1. The bombarding electrons can transfer a very wide range of electronic energy to the molecule. When only a small energy is transferred, insufficient to expel an electron from the molecule, excitation rather than ionization occurs. Kinetic energy is lost by the bombarding electron in an amount equal to the energy transferred to the molecule. Because the range of allowed electron kinetic energies is effectively continuous, the exact amount of excess energy over that required to form an ion in any particular electronically excited state can be removed by the

---

*The term "parent ions" has been used in the past interchangeably with "molecular ions", signifying that these ions are the parents of all the ions making up the mass spectrum. We shall not reserve the term "parent ion" for such a meaning but will use it to describe any fragmenting ion; the product ion formed from a fragmenting parent will be referred to as a daughter ion.

ejected electron. As already mentioned in Chapter 1, the time of interaction between a bombarding electron and a molecule is very short and
ionization takes place by a "vertical" process. That is to say, the
electron energy measured as the minimum necessary to produce
ionization is the energy necessary to remove an electron keeping all
nuclei in the positions corresponding to their instantaneous positions
in the neutral molecule.

When large amounts of energy are transferred, sufficient to expel
an electron from the molecule, a molecular ion is formed. The
probability that this will occur is a function of the kinetic energy of
the bombarding electron. A plot of this probability as a function of
electron energy is called the ionization efficiency curve. The curve
for bombardment of mercury vapor is shown in Fig. 10. It can be
seen that at the ionization potential, the probability of ion formation
is zero, but rises as the ionizing electron energy is increased.

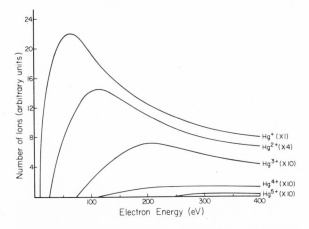

Fig. 10. Ionization efficiency curves for mercury. The curves show the probabilities of producing mercury ions carrying various numbers of positive charges
as a function of the energy of the bombarding electrons.

At, and just above the ionization potential only singly charged molecular ions are formed, but at higher energies other ions are observed
(see below). It should be noted that the minimum energy required to
effect ionization by electron impact, the vertical ionization potential,
may be greater than the adiabatic ionization potential. The latter is
defined as the energy difference between the ground vibrational
levels of the lowest electronic states of the molecule and molecular
ion.

MULTIPLY CHARGED IONS

If sufficient energy is transferred from the bombarding electron to the molecule, two or more electrons can be removed, resulting in ions carrying multiple charges. The processes for formation of doubly and triply charged molecular ions by electron bombardment can be represented by

$$M + e \rightarrow M^{2+} + 3\,e$$

and

$$M + e \rightarrow M^{3+\cdot} + 4\,e$$

The minimum bombarding electron energies for formation of mercury ions carrying 2 and 3 charges can be read from Fig. 10. In general, the energy necessary to remove successive electrons increases as more electrons are removed. The total energy necessary to remove one, two, three, four, five and six electrons from neutral argon is 15.8, 43, 84, 144, 219 and 310 V, respectively [114].

Multiply charged ions can also be formed from polyatomic molecules and ions carrying two and three positive charges are often observed in the mass spectra of organic compounds. In mass spectra, ions are separated according to their mass-to-charge ratios, so that multiply charged ions can be recognized by the occurrence of apparent mass intervals of one-half or one-third of a mass number in the mass spectra.

FRAGMENT IONS

The quasi-equilibrium theory (QET) of mass spectra (see Appendix I) explains how the molecular ions that contain excess internal energy (initially in the form of electronic excitation energy) above the ground state can decompose to form fragment ions and neutral fragments. Either moiety can contain an odd or an even number of electrons. The process can be represented by

$$M^{+\cdot} \rightarrow F_1^{\,+} + N_1^{\,\cdot}$$

when the daughter ion is an even-electron species, or by

$$M^{+\cdot} \rightarrow F^{+\cdot} + N$$

when the ion is an odd-electron species. The quasi-equilibrium theory also explains how the molecular ions can decompose in more than one way and how a series of competing unimolecular reactions can occur leading to several different fragment ions. We may represent the situation by saying that reactions such as

$$M^{+\cdot} \rightarrow F_2^+ + N_2{}^\cdot$$

and

$$M^{+\cdot} \rightarrow F_3^+ + N_3{}^\cdot$$

can also occur. The mechanism by which the excess electronic energy becomes converted into vibrational energy in a series of unimolecular radiationless transitions between crossing potential energy surfaces is discussed in greater detail in Appendix I.

Cleavage of a particular bond in a molecular ion leads to fragments F_1 and N_1 and the reaction has been illustrated above with the charge located in the fragment F_1. In addition to this most probable reaction there may also be a complementary reaction leading to $F_1{}^\cdot$ and $N_1{}^+$. A useful rule regarding the location of the positive charge[130] was first formulated by Stevenson[252] who noted that for hydrocarbons, the positive charge tended to remain on the fragment with the lower ionization potential. Stevenson's rule can be simply explained by using the energy diagram shown in Fig. 11. This shows

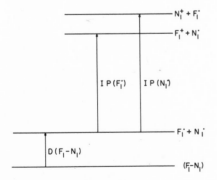

Fig. 11. Illustration of Stevenson's Rule predicting that for the fragmentation of a positive ion, the radical of lower ionization potential is more likely to carry the positive charge.

the energy necessary to produce $F_1{}^+$ and $N_1{}^{\cdot}$ as compared to $F_1{}^{\cdot}$ and $N_1{}^+$. The appearance potential of $F_1{}^+$ (the sum of the dissociation energy of the $F_1{-}N_1$ bond and the ionization potential of $F_1{}^{\cdot}$) is lower than that of $N_1{}^+$. Excess internal energies of the fragments are ignored in this simple treatment. Such processes are discussed in more detail in Chapter 4. It will be seen later that the rate of any reaction will increase as the excess energy contained in the ion above the minimum amount necessary for reaction increases. Thus, all other things being equal, the reaction leading to the products of lowest energy will predominate.

The fragment ion $F_1{}^+$ formed in a single step from the molecular ion is known as a primary fragment ion. The primary fragment ion may be formed in a vibrationally excited state and react further. Thus, we may write

$$F_1{}^+ \rightarrow X_1{}^+ + Y_1$$

or less probably, since electron unpairing is involved,

$$F_1{}^+ \rightarrow X_2{}^{+\cdot} + Y_2{}^{\cdot}$$

The ions $X_1{}^+$ and $X_2{}^{+\circ}$ are known as secondary fragment ions. Obviously, such processes can continue and give tertiary and further products. The situation may be illustrated with the aid of the energy diagram in Fig. 12. The appearance potential of $F_1{}^+$ is the minimum

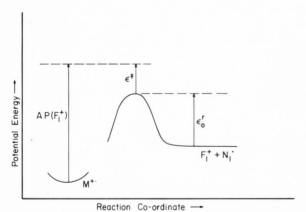

Fig. 12. Excess energy involved in measurements of the appearance potentials of fragment ions. Internal energy, $\epsilon^{\ddagger}$, is necessary in the activated complex for the reaction to occur within the time scale appropriate to the mass spectrometer. A reverse activation energy, $\epsilon_0{}^r$, may also be present and may lead to formation of products containing internal or translational energy. Zero point energies have not been illustrated in this simple figure.

energy leading to observation of $F_1{}^+$ ions (*i.e.* forming $F_1{}^+$ ions in the source in a time less than 10^{-6} sec) and the activated complex may contain some excess (vibrational) energy $\epsilon^{\ddagger}$. Reaction to form $F_1{}^+$ and $N_1{}^{\cdot}$ in their ground vibrational states could release the total amount of energy $\epsilon_0^r + \epsilon^{\ddagger}$ (necessarily as translational energy of the products) but a proportion of this energy may, alternatively, be shared between the products as vibrational energy so that $F_1{}^+$ may be able to react further. This is discussed in more detail in Chapter 4. The primary fragmentation of the molecular ion discussed above has been depicted as leading to an even-electron ion $F_1{}^+$ and an odd-electron neutral fragment. The secondary fragmentations have been depicted as leading either to an even-electron ion $X_1{}^+$ and a neutral molecule Y_1 (even electron) or a positively charged free radical $X_2{}^{+\cdot}$ and a free radical $Y_2{}^{\cdot}$ (odd electron). Both processes are common in the fragmentation of organic ions, but because the even-electron products containing only paired electrons tend to be the more stable, the reactions leading to these products are usually predominant. Fragment ions can also be formed by the process of "pair production" in which both a positively and negatively charged fragment arise simultaneously without the intermediate formation of a positively charged molecular ion. The process can be written

$$M + e \rightarrow F^+ + G^- + e$$

Special equipment has been designed [254] for detecting and studying simultaneous formation of a positive ion and a negative ion in processes such as the above. Pair production is most likely to occur when both positive and negative ions are particularly stable. The process has been noted [200] in halogenated compounds that can give stable atomic halogen negative ions.

REARRANGEMENT IONS

Often, the fragmentation of an organic ion involves a transfer or rearrangement of the atoms between the fragments during or before breaking of the bond. Thus, we may write

$$ABCD^{+\cdot} \rightarrow AD^{+\cdot} + BC$$

Commonly, atoms such as hydrogen or fluorine are transferred, but rearrangements involving much larger groups such as $CH_3{}^{\cdot}$ or $C_6H_5{}^{\cdot}$ are also known. It is a general rule that rearrangements tend to lead to products of high stability, *i.e.* for which $\Delta H_f^\circ(AD^{+\cdot}) + \Delta H_f^\circ(BC)$ is low. This is necessary to compensate for the low frequency factors of these reactions. That is, for rearrangement reactions to occur at rates which result in observable amounts of products, their unfavorable entropies of activation must be balanced by particularly favorable activation energies. The low activation energies of many rearrangement reactions account for the fact that abundant unimolecular metastable ions are frequently associated with rearrangement reactions.

In saturated carbonium ions of formula $C_nH_{(2n+1)}^+$, a rearrangement into an even-electron daughter ion and even-electron neutral molecule gives the most stable products. For unbranched long-chain ions, several reactions give products of almost equal energy and a variety of daughter ions ensues. For example, the reactions

$$C_8H_{17}^+ \quad
\begin{cases}
\longrightarrow C_6H_{13}^+ + C_2H_4 \\
\longrightarrow C_5H_{11}^+ + C_3H_6 \\
\longrightarrow C_4H_9^+ + C_4H_8
\end{cases}$$

all occur readily. The variety of products formed led to the term "random" rearrangements [185] to describe these reactions.

In other cases, for example, where the ion contains a heteroatom such as oxygen, one, or at most a few, rearrangements will give rise to abundant product ions. Fragmentation is visualized as occurring from ion structures in which the charge is localized on the heteroatom or in some other location that gives a ground-state ion of low ΔH_f°. This is understandable because for an ion of a particular total energy content, such states allow the maximum amount of energy to exist in the vibrational form. Consider, for example, the McLafferty rearrangement of ketones which is usually written

The half-arrow or "fish-hook" [73] is used to denote the movement of a single electron. Hydrogen rearrangements generally occur *via* five-, six-, or seven-membered cyclic transition states. The formation of the neutral olefin in the reaction confers stability upon the products. It should be noted that in the product ion shown, O^+ is trivalent, one of its lone-pair electrons having been removed in the ionization process. A similar increase in valency upon ionization can occur in a nitrogen atom and the controlling influence of product stability can be seen, for example, in the formation of the ammonium ion $R\overset{+}{N}H_3$ from secondary amines.

Skeletal rearrangements are those in which bond formation involves other than monovalent atoms. The reactions which occur are again governed by the considerations of product stability, so that the elimination of stable neutral molecules from within an ion is common.

A series of consecutive rearrangement reactions may result in the statistical distribution of a number of atoms or groups between different positions in the ion. As a result of this process, an isotopic label such as deuterium may be statistically "scrambled" within the ion. With a small number of rearrangement reactions this scrambling may be incomplete. These reactions are merely a particular type of reactant ion isomerization; they are often characterized by extremely low activation energies.

It is usual to distinguish between "tight" and "loose" transition complexes from which rearranged fragments are formed. A "tight" complex is one in which considerable deformation of bonds has occurred resulting in hindered rotation and a decrease in the number of rotational degrees of freedom. Formation of this complex has a high entropy of activation and a low frequency factor. In contrast, formation of a "loose" complex requires a low entropy of activation and a higher frequency factor. In some cases, rearrangement involves movement of large groups over relatively large distances so that the time needed to form the transition state may be long even though the associated activation energy may not be high. In such cases, the frequency of formation of states reacting with high rate constants will be expected to be small.

STABLE IONS

A stable ion is defined as one that is formed within the ionization chamber and travels from the ionization chamber to the collector without decomposition. The time taken for an ion to travel over this distance will depend upon the potential used to accelerate the ion, upon its mass-to-charge ratio and upon the dimensions of the mass spectrometer. For a large instrument such as the double-focusing MS9 made by AEI/GEC[93], using an accelerating voltage of 8 kV, an ion of mass-to-charge ratio 100 will reach the collector about 10^{-5} sec after leaving the ionization chamber. The time of flight of an ion through the instrument after acceleration is proportional to

$(m/xeV)^{\frac{1}{2}}$ where m is the mass, x the number of charges and $\mathring{V}$ the accelerating voltage. The denominator of this expression corresponds to the kinetic energy of the ion, a property with which we shall be greatly concerned. It should be emphasized that the kinetic energy acquired by an ion during acceleration is proportional to the number of charges carried by the ion. The time of flight through a mass spectrometer can be varied if electrode potentials are changed so that ions move more slowly than normally over part of their path. This can be accomplished in some instruments such as the RMH-2 made by Hitachi [44] which is fitted with a lens system in which an ion can be maintained at a constant, low kinetic energy for several microseconds.

The normal peaks in a mass spectrum are due almost entirely to stable ions, although they do contain a small contribution from ions that have decomposed in the field-free region immediately in front of the collector. Obviously, the ion species that are counted as stable will depend upon the range of time after ionization over which the spectrum is measured. Instruments of different size operated under different conditions will give somewhat different mass spectra.

UNSTABLE IONS

An unstable ion is one that is formed with sufficient internal energy that it decomposes before leaving the ionization chamber, *i.e.* in a time of the order of 10^{-6} sec. Decomposition will continue until the products contain little excess energy above their ground states and are stable or metastable (see below). The distribution of internal energies that can be produced in an organic molecule by electron bombardment depends upon the available energy possessed by the bombarding electron. The mass spectrum of singly charged ions becomes insensitive to changes in electron energy above about 30 eV suggesting that the distribution of excess energies given to the molecule is changing only slowly. At lower electron energies, especially within a few volts of the first ionization potential, the mass spectrum changes dramatically with electron energy. As the appearance potentials of various fragment ions are exceeded, the molecules are given sufficient energy so that more and more molecular ions react with large enough rate constants to produce ions that are themselves unstable.

METASTABLE IONS

A metastable ion is one that is sufficiently stable to leave the ionization chamber, but that decomposes before reaching the collector. Many ions that decompose in regions in which there is an electric or magnetic field collide with the walls of the instrument and do not reach the collector. Others that do reach the collector are spread over a wide range of apparent mass-to-charge ratios and are difficult to detect; these are discussed in Chapter 3. But ions that decompose in the field-free region in front of the magnetic sector will be brought to a focus and give rise to the so-called metastable peaks. The focusing of the daughter ions from metastable parents, the positions of the metastable peak on the mass scale and the shapes of these peaks are all discussed in Chapter 3. Ions that decompose in front of the electric sector of a double-focusing mass spectrometer will be unable to pass through this sector because they will have lost kinetic energy. It is described later how, if the accelerating voltage or electric sector voltage is changed by the appropriate amount, such ions can also be focused and collected.

An ion will thus be deemed to be metastable and give rise to a metastable peak if it decomposes in a field-free region. The time taken to reach these regions is about one-third to two-thirds of the time taken by an ion to reach the final collector as discussed above. For the decomposition of the metastable ion to be observed, it will be necessary that a significant number of them be present with the appropriate amount of internal energy for the corresponding rate constant to lie between 10^4 and 10^6 sec^{-1}. Usually, if high-energy bombarding electrons of, say, 70 or 100 eV have been used, molecular ions will be formed with a broad spectrum of internal energies. This will include energies corresponding to the mean fragmentation rate appropriate for observation of the metastable ion transition. In the case of fragment ions, too, these will have been formed by way of processes with a large range of rate constants and they are likely to include ions with an appropriate residual internal energy for observation of the metastable peaks appropriate to their decomposition. A metastable transition corresponding to a particular process may still, however, not take place for one of several reasons discussed later in Chapter 4 and Appendix I. They include the following: (i) the transition complex for the particular reaction may be too "tight" a complex leading to a low-frequency factor for the decomposition; (ii) there may be a second fragmentation route giving a product of considerably lower appearance potential. In this case a different metastable transition might be observed, the rate constant for this reaction increasing with energy in such a way that it becomes so fast that essentially all ions with sufficient energy to undergo the second reaction have been removed by the time the field-free region is reached. (*iii*) Alternatively, it might be necessary for the parent ion to undergo an internal rearrangement involving actual transfer of atoms between different parts of the molecule, without loss of mass, before the metastable transition can occur. In these cases the rate of fragmentation may depend on the rate of formation of the rearranged species which may not, strictly, be metastable in the sense of having a half-life after formation of the order of a few microseconds. We shall come across other cases of reactions that give rise to peaks indistinguishable from metastable peaks in which the fragmenting ion is not metastable. Another noteworthy case, especially in small ions, occurs when fragmentation occurs by a predissociation (curve-crossing) mechanism. Normally, predissociation, which involves a transition from one potential energy hypersurface to another, occurs rapidly and the products are formed in the ion chamber. However, if curve crossing is forbidden because of symmetry or spin considerations, some ions will nevertheless undergo the transition but will do so at a reduced rate. This may result in fragmentation in the field-free region.

Often, decomposition of a stable ion in the appropriate field-free region may be induced by collision with a neutral species. We shall consider these and other reactions brought about by collision gases as lying within the scope of this book; some of the processes are discussed in later sections.

ION—MOLECULE REACTIONS

The interaction of ions and molecules can lead to a variety of processes that are important in giving us a better understanding of the structures of the ions themselves and the mechanisms and kinetics involved in their reactions. Ions of low kinetic energy form collision complexes with neutral molecules within the ionization chamber. Fast-moving beams of ions can be caused to fragment by collision with gas molecules and various processes that involve changes in the number of charges carried by the ion can also occur by interaction with a collision gas. An understanding of the structures of collision complexes and of the various mechanisms by which transfer of energy can take place is necessary for an understanding of the reactions of metastable ions. Hence, the various types of interaction of molecules with ions will be detailed below. Our concern in this book is mainly with reactions of high-kinetic-energy ions. For completeness, brief reference is made below to ion—molecule reactions that occur at low kinetic energies, for example, in the ion source.

COLLISION-INDUCED FRAGMENTATION

It is tacitly assumed by many workers in mass spectrometry that the presence of broad, diffuse peaks in a mass spectrum is certain evidence that fragmentations of metastable ions are being observed. Although such peaks are indicative of reactions occurring in the field-free regions of the mass spectrometer, there is no *a priori* reason why they should be due to unimolecular decompositions and it may well be that the majority of the weak peaks previously identified as metastable peaks have, in fact, been due to reactions induced by collisions with residual gas molecules in the mass spectrometer. There are two ways of gaining information on this point. The first is to vary the pressure of background gas in the appropriate field-free region in a controlled fashion and, if the height of the observed peak increases with gas pressure, to plot peak height against pressure on a logarithmic scale and to estimate the pressure that could correspond to the initially observed peak height. If this pressure is of the order of the residual pressure, it could explain the presence of the peak. The second method is to measure carefully the shape of the observed peak as the background pressure is changed. If the shape of the peak alters as background gas is introduced, this is a good indication that the original peak observed was, indeed, due to a unimolecular process and that there is a corresponding collision-induced reaction. It is shown in Chapter 3 that the width of a metastable peak depends only upon the internal energy released in the fragmentation of the metastable ion. This energy release is characteristic of the reaction and if it changes with pressure it is an indication that a new reaction has been induced by the collision gas.

When fragmentation of an ion occurs unimolecularly, the kinetic energy of the ion is shared between the fragments in proportion to their masses. In other words, the mean velocity of the fragments is the same as that of the metastable ion and the position of the metastable peak on the mass scale may be calculated to a high degree of accuracy on this assumption. When the same fragmentation is induced by collision, the position on the mass scale of the resultant peak is very slightly changed showing that in the collision, a perceptible fraction of the kinetic energy of the bombarding ion has been lost to internal energy. This is discussed further in Chapter 4. If the pressure of collision gas is increased to such an extent that collisions involving transfer of momentum become appreciable, the observed peak heights begin to decrease as ions are scattered out of the beam and lost. It is not collisions of this kind involving transfer of momentum between the ion and the nuclei of the collision gas that cause the small but measurable displacement of the metastable peak on the mass scale. The collisions that are effective in inducing the observed fragmentation are due entirely to interactions between the ion and electrons of the collision gas. In these collisions, ion kinetic energy, the only available energy possessed by an ion in its ground state, is converted directly into electronic excitation energy in the ion itself. In order for the resultant fragmentation to take place as the ion is passing through the field-free region, the rate constant must be of the order of 10^6 sec^{-1} or greater. The necessary excess energy for this to occur is of the same order as that producing fragmentations in the ionization chamber. The observed collision-induced peaks can therefore give direct information concerning the pathways leading to formation of the normal ions that make up the conventional mass spectrum. The abundances of the observed peaks should correlate with the abundances of corresponding peaks in the mass spectrum. Such a correlation has indeed been noted by Jennings[154] in the case of benzene. The effective cross-section for these collisions is high, being of the order of hundreds of square Ångstrom units. Careful control and measurement of collision gas pressure is necessary to obtain the maximum observable effects. Typically, for a collision region 500 mm long, a collision gas pressure of the order 10^{-3} Pa ($\sim 10^{-5}$ torr) is optimum. The majority of the collisions will, presumably, involve the outer molecular orbitals and thus an energy transfer of less than about 10 eV.

CHARGE-EXCHANGE COLLISIONS

If an ion carrying two positive charges passes through a collision gas, N, a charge exchange reaction can take place in which the doubly charged ions lose one of their charges and the collision gas molecules become ionized. The reaction may be written

$$M^{2+} + N \rightarrow M^{+\cdot} + N^{+\cdot}$$

If the ion M^{2+} is in its ground vibrational state, the reaction can proceed without change in the kinetic energy of the ion if the standard heats of formation of the products are together less than those of the reactants. This will be true if the ionization potential of N, IP(N), is smaller than the ionization potential of $M^{+\cdot}$, IP($M^{+\cdot}$). This

can be seen by reference to Fig. 13. It is possible for kinetic energy to
be converted into electronic energy (see next section), so that the
above energetic considerations do not constitute a necessary condition
for the reaction to proceed. The efficiency of the process is certainly
much increased, however, if the above condition is met and if excited
states are available in $M^{+\cdot}$ or N so that there can be a perfect match

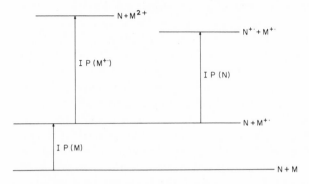

Fig. 13. Energetic considerations relevant to charge exchange reactions between
a doubly charged ion and a neutral molecule.

between the energy that is available and that of the products. The
effective cross-section for charge exchange is of the same order of
magnitude as for collision-induced fragmentation; a pressure of the
order of 10^{-3} Pa (10^{-5} torr) will, typically, give the most
intense peak due to product ions. All doubly charged ions, of whatever mass, acquire during acceleration twice the kinetic energy of
singly charged ions. To a first approximation, they suffer no loss of
energy during charge exchange and so the product ions have twice
the kinetic energy of all normal singly charged ions and can be separated from them. It has been claimed [264] that, for certain classes
of organic compound, the mass spectrum of the doubly charged ions
is more useful for diagnosing structural features than the normal
mass spectrum. Charge exchange offers a convenient method of ob-
taining such spectra. Further features of these spectra are discussed
in Chapter 5. Ions containing more than two charges can also undergo charge exchange.

IONIZATION BY THE COLLISION GAS

It is known that when singly charged ions of high kinetic energy
travel through a collision gas, they can be ionized and converted into
doubly charged ions. When this occurs, the total energy of the
system must remain constant. The internal energy is increased by

removal of an electron and the only available source of this energy is the kinetic energy of the ion. But when the kinetic energy of the ion is reduced, its linear momentum will also be reduced and the law of conservation of momentum therefore dictates that some linear momentum (and hence kinetic energy) be given to the molecules of the collision gas. In addition, in a "glancing" collision both products of the collision may acquire angular momentum. Further discussion of these points is given in Chapter 4. The effect of ignoring the momentum of the collision gas molecules does not lead to serious error except in the case where the mass of the bombarding ion greatly exceeds that of the collision gas. An electric sector transmits ions of a particular kinetic energy to charge ratio. The kinetic energy of the bombarding ions is reduced slightly in the collision and the resultant doubly charged ions will be transmitted at a value of the sector voltage just less than half of the normal value. The difference between the actual energy of the transmitted ion and half the energy of the bombarding ions is a measure of the energy needed to ionize the singly charged ions and is, to a first approximation, independent of the nature of the collision gas used.

ION—MOLECULE REACTIONS WITHIN THE IONIZATION CHAMBER

If the sample pressure is raised within the ionization chamber, peaks can often be observed at mass-to-charge ratios greater than that for the molecular ion itself. Such peaks are particularly common in the case of molecules containing polar groups. When ionization occurs by the removal of one of the lone-pair electrons from an oxygen or nitrogen atom, a strong bond can be formed by radical attachment at this site. The most commonly observed ion—molecule reaction product corresponds to the attachment of a hydrogen atom to give a peak corresponding to the ion $(M + 1)^+$. However, cases have been known for some time where larger groups such as acetyl can be transferred to the molecular ions[34]. The probability of a collision leading to reaction increases with the pressure within the ionization chamber so that it is usual to work at as high a sample pressure as possible. With specially designed ionization chambers capable of being operated at pressures as high as several atmospheres, many new ionic species have been studied.

The very intense fields near to the sharp edge used in field ionization [21,150] have the effect of attracting polar molecules so that the local pressure of a polar compound can be high in the region in which ionization is effected, even though the mean pressure in the source is low. Indeed, with such a source it is difficult to avoid the effects of reactions of the sample ions with water and ions such as $[M + (H_2O)_n]^{+\cdot}$ are common in the mass spectra of organic molecules as are ions such as $(M_2 + H)^{+\cdot}$. A detailed account of this special field is outside the scope of the present book, however.

Modified ionization chambers are available commercially which allow sample pressures up to the order of 100 Pa (about 1 torr) to be used. These sources differ from those usually used in mass spectrometry in being more "gas tight". Necessary holes for admission of ionizing electrons and for extraction of ions are small and good differential pumping of the source and analyzer regions of the mass spectrometer is necessary due to the high pressures employed. At the highest pressures, penetration of the electron beam into the ionization chamber is limited unless the electron energy is increased to several hundred electron volts. The number of ions that can be drawn from an ionization chamber is limited and the effect of increasing the pressure is, at first, to reduce the sensitivity in terms of the ion current per unit of pressure. As the pressure is further increased, however, more ions are extracted by viscous flow of sample from the source. High-pressure sources are used in the technique of chemical ionization [109,213]. A small amount of sample (0.01—0.1%) is introduced into the ionization chamber together with an excess of a reactant gas, usually a hydrocarbon. Under these conditions, it is much more likely that a sample molecule will react with an ion originating from the hydrocarbon than be ionized directly by an electron. If, for example, methane is used as the reactant gas, electron bombardment produces the primary ions $CH_4^{+\cdot}$ and CH_3^+. These react with methane molecules to produce mainly CH_5^+ and $C_2H_5^+$ ions. These are the ions largely responsible for ionization of the sample molecules by proton transfer and hydride abstraction reactions, respectively. Sample ions produced in this way, together with those formed by subsequent fragmentation, make up the chemical ionization spectrum. The ion—molecule ionization process involves slow, equilibrium changes of electronic states and the positions of nuclei. Ionization is not a vertical process like that produced by electron bombardment and so the chemical ionization spectra are different from electron-impact spectra. They are usually characterized by a prominent peak corresponding to $(M + 1)^+$ and/or $(M-1)^+$ ions. In some cases [108] $(M + 29)^+$ and $(M + 41)^+$ ions are also seen and it is necessary to have some preliminary knowledge of the type of sample with which one is dealing before meaningful structural interpretations can be made. The method has proved useful in giving prominent ions in the molecular ion region for many compounds that give only very small molecular ion peaks under electron bombardment.

The ionization chambers designed for chemical ionization work can be used to advantage in the general study of ion—molecule reactions. A pressure of one torr of sample molecules enables many ion—molecule reaction products to be identified. Even without the use of specially-designed ionization chambers, it is possible to detect ion—molecule reactions in a high-sensitivity instrument [45,51]. A study of these low-energy ion—molecule reactions will not form a major theme in this book; in the study of collisions between ions and molecules, we shall be mainly concerned with the various reactions that result when ions of high kinetic energy (typically, thousands of electron volts) are used. The two fields are largely complementary and each can give unique information about the energetics and mechanisms of the processes involved.

Chapter 3

Focusing and kinetic energy measurement of ion beams

"Get you home, you fragments"

Shakespeare, *Coriolanus*

Consider a mono-energetic beam of ions issuing from a slit and entering a field-free region with a divergence angle 2α as shown in Fig. 14. Suppose that the beam enters either a sector magnetic field or a radial electric field that focuses the ion beam on to a collector slit. Suppose that an ion $m_1{}^+$ decomposes in the field-free region,

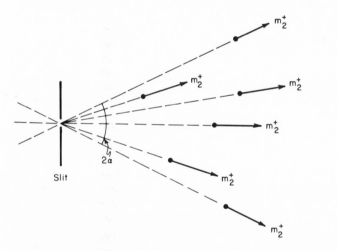

Fig. 14. Fragmentation of mono-energetic ions in a field-free region. No kinetic energy is released and there is no change in velocity or direction of the fragments as compared with the reactant ions.

yielding a daughter ion $m_2{}^+$ and a neutral fragment m_3. If this decomposition takes place without any conversion of internal (excitational) energy into external (kinetic) energy, each daughter ion $m_2{}^+$ will continue to move along the original direction of motion of its parent ion $m_1{}^+$ and with the same velocity. There are several important consequences that result from this simple fact and it is worthwhile to emphasize them at this stage. Energy will, of course, be conserved in the fragmentation; the initial kinetic energy $\frac{1}{2}m_1v^2$ will be equal to the total kinetic energy ($\frac{1}{2}m_2v^2 + \frac{1}{2}m_3v^2$) of the fragments and it can readily be seen that this kinetic energy is shared between the fragments in the ratio of their masses. Since the directions of motion of all the daughter ions coincide with the original directions of motion of the ions m_1^+, they will appear to be issuing from the slit as a mono-energetic beam with the same divergence angle as the original parent ions. This is illustrated in Fig. 14. Now, such a beam can be focused by the sector magnetic or electric field to give an image indistinguishable from that obtained from a beam of stable ions m_2^+ issuing from the same slit. Thus, as the magnetic or electric field is scanned, a peak will be seen due to the m_2^+ ions from the decomposition of the metastable m_1^+ ions. The peak obtained using a magnetic sector is known as a metastable peak; when obtained using an electric sector it can be called an ion kinetic energy (IKE) peak. It should be noted that the peak observed will be the same shape as the corresponding peak due to stable ions.

METASTABLE PEAKS IN NORMAL MASS SPECTRA

In double-focusing mass spectrometers of Nier—Johnson geometry, the electric sector precedes the magnetic sector. The daughter ions from the decomposition of metastable ions in the field-free region in front of the electric sector will not be seen in conventional mass spectra. This field-free region is, conventionally, known as the first field-free region. The electric sector voltage is coupled to the main accelerating voltage so that only ions having the full acceleration energy will be transmitted through the electric sector. Ions m_2^+, formed as described above, will strike the electric sector walls and be lost. Products formed from ions that decompose in the field-free region in front of the magnetic sector (conventionally known as the second field-free region) will give rise to peaks at the final collector of the mass spectrometer as the magnetic field is scanned. The position of such peaks on the mass scale can readily be determined. The velocity of the metastable m_1^+ ions entering the second field-free region is given by eqn. (4) as

$$v = \left(\frac{2eV}{m_1}\right)^{1/2}$$

After the metastable transition, the momentum of the product m_2^+ ions will be given by

$$m_2 v = m_2 \left(\frac{2eV}{m_1} \right)^{1/2}$$

The radius followed by these ions in the magnetic sector is given by eqn. (3) as

$$r = \frac{m_2 v}{Be} = \frac{m_2}{Be} \left(\frac{2eV}{m_1} \right)^{1/2}$$

Thus

$$\frac{m_2^2}{m_1} \frac{1}{e} = \frac{m^*}{e} = \frac{B^2 r^2}{2V} \tag{16}$$

Comparing eqn. (16) with eqn. (5), it can be seen that the daughter m_2^+ ions are transmitted through the magnetic sector with an apparent mass m^* (ref. 140) where

$$m^* = \frac{m_2^2}{m_1} \tag{17}$$

The same metastable peaks, due to decompositions in front of the magnetic sector, will also be seen in single-focusing sector instruments. An example of metastable peaks is shown in Fig. 15. The apparent masses m^* will, in general, not correspond to integral mass values. However, m_1 and m_2 will be integers and the values of m_1

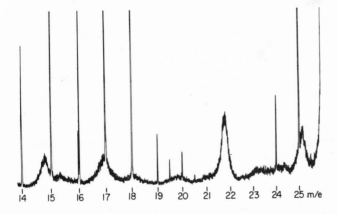

Fig. 15. Part of a mass spectrum between mass-to-charge ratios 14 and 26 showing a variety of diffuse (metastable) peaks.

and m_2 can often be determined uniquely (and in the worst case reduced to a few possible values) by reference to published tables[35] that list the values of m^* corresponding to all integral values of m_1 and m_2 up to 500.

ION KINETIC ENERGY SPECTRA

In order to detect decomposition occurring in the first field-free region, it is necessary to lower the voltage applied between the plates of the electric sector relative to the accelerating voltage. It can be seen from eqn. (10) that the kinetic energy-to-charge ratio of the ions transmitted along the central path through the electric sector is directly proportional to the field strength, which is itself proportional to the voltage applied across the plates. If the voltage across the electric sector is changed to a fraction m_2/m_1 of the value at which ions that carry the full energy of acceleration are transmitted (the so-called main beam of ions), then the daughter $m_2{}^+$ ions will pass along the central path. If E_1 is the voltage at which the daughter ions of the metastable transition are seen and E is the voltage corresponding to the main beam, then

$$E_1 = \frac{m_2 E}{m_1} \tag{18}$$

The signal obtained on a collector electrode placed behind the β-slit (located in the energy focus plane after the electric sector) as the electric sector voltage is scanned, is called an ion kinetic energy (IKE) spectrum[47, 48, 162]. A typical IKE spectrum is shown in Fig. 16. An electric sector separates a beam of ions according to their kinetic energies. For a mono-energetic beam containing several different ionic species, there is, thus, no mass separation. The relationship of eqn. (18), for a particular small range of values of E_1, may be satisfied by more than one pair of values m_1 and m_2. There thus may be overlap of peaks in the IKE spectrum. Overlapping may occur anywhere in the spectrum, but is particularly likely in the region where $E_1 \cong E$, close to the main beam. Most organic compounds form fragment ions by loss of one or more hydrogen atoms from their molecular ions; IKE peaks at voltage $[(m_1-1)/m_1]E$, $[(m_1-2)/m_1]E$, $[(m_1-2)/(m_1-1)]E$, etc. will lie very close together. It is, of course, possible to mass analyze the ions giving rise to any peak in the IKE spectrum. The peak is focused on the intermediate collector located behind the β-slit (see Fig. 45, p. 83) and then the

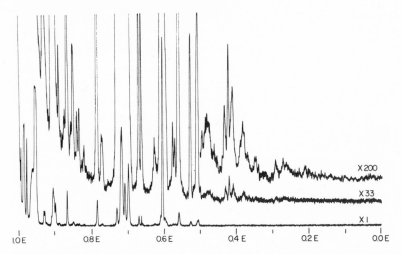

Fig. 16. The ion kinetic energy (IKE) spectrum of *n*-heptane from an electric sector voltage at which the main beam of ions is transmitted (E) to zero. Overlapping peaks can be seen, particularly in the region of the main beam.

collector is raised out of the beam, allowing the ions to pass into the magnetic sector for mass analysis. The mass spectrum thus produced will consist of a single mass peak if the IKE peak was the result of a single reaction.

SPECTRA OBTAINED BY ACCELERATING VOLTAGE SCANNING

There is an alternative method by which ions that have lost kinetic energy in metastable transitions in front of the electric sector can be transmitted through this sector. This is to keep the electric sector voltage fixed at the normal value E but to change the main accelerating voltage from its normal value V to a new value V_1. The voltage E across the electric sector plates is appropriate for transmitting stable ions that have received the full, normal accelerating energy, eV. Thus, in order for any ion to be transmitted through the electric sector, it is merely necessary to bring it to a kinetic energy equal to eV by increasing the accelerating voltage. Suppose that an ion m_1^+ falls through the new accelerating potential V_1. Then the daughter ion, m_2^+, will continue its motion with a fraction of this kinetic energy, $(m_2/m_1)eV_1$. If $(m_2/m_1)eV_1 = eV$, i.e. if

$$\frac{m_2}{m_1} V_1 = V \qquad\qquad (19)$$

the daughter ion will be transmitted through the electric sector and can be detected on a collector placed behind the β-slit. It is obvious that a spectrum similar to an IKE spectrum can be obtained by scanning the accelerating voltage and that overlapping of peaks might be expected in this method, also. It is, however, possible to avoid this difficulty and at the same time to obtain additional information by making use of the magnetic sector that follows the electric sector in conventional double-focusing instruments. This was first done by Jennings[152], by Futrell et al.[117] and by Barber and Elliott[15]. To understand their method, consider the mass spectrometer to be set up so that stable ions $m_2{}^+$ are transmitted through the electric sector with kinetic energy eV and that the magnet current is adjusted so that these $m_2{}^+$ ions are transmitted through the magnetic sector and collected. Keeping the current through the magnet and the voltage across the electric sector fixed, the main accelerating voltage is now scanned upwards to the new value V_1. At this value, we have seen that $m_2{}^+$ ions formed as daughters of metastable $m_1{}^+$ parents will acquire a kinetic energy eV. They will thus be transmitted through both the electric and magnetic sectors and be collected. It can be seen that as V is scanned, all daughter ions of mass m_2 formed from various parents will be collected in turn, the ratio of the new to normal accelerating voltage at which a peak is seen giving the ratio of parent-to-daughter mass. A high voltage (HV) scan showing several routes along which ions of a particular mass are formed is shown in Fig. 17.

ENERGY SPECTRA OBTAINED WITH PRIOR MASS ANALYSIS

In order to obtain unequivocal information concerning ion fragmentations from an electric sector, and to avoid the difficulties due to overlapping of IKE peaks, it is desirable to carry out mass separation of the ion beam prior to energy analysis. This can readily be accomplished by reversing the positions of the magnetic and electric sectors in a double-focusing mass spectrometer so that the ion beam passes through the magnetic field first[54, 56]. An ion of any mass-to-charge ratio can then be isolated by the magnetic field and its subsequent fragmentations studied by scanning the electric sector. The mass-analyzed ion kinetic energy spectra obtained in this way are referred to as MIKE spectra and that for the molecular ion of n-heptane is shown in Fig. 18. Every peak in this spectrum occurs in the IKE spectrum of n-heptane (Fig. 16) which, however, includes numerous other peaks due to fragmentations of metastable ions

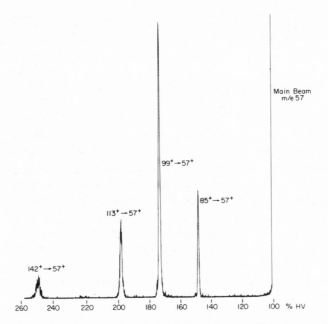

Fig. 17. HV scan of m/e 57 from n-decane. The fragment ions can be seen to arise from ions of mass-to-charge ratios 142, 113, 99 and 85.

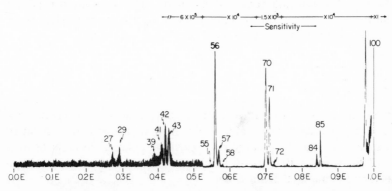

Fig. 18. Mass analyzed ion kinetic energy (MIKE) spectrum of n-heptane. The spectrum shows the fragmentation products of the molecular ion, m/e 100. Note the narrowness of the molecular ion signal with respect to the metastable peaks.

other than the molecular ion. The design requirements for a MIKE spectrometer, or IKE spectrometer with mass analysis of the product ions, are discussed later.

CALIBRATION OF THE MASS SCALE FOR VARIOUS METHODS OF SCANNING

As discussed in Chapter 1, a magnetic sector measures mass-to-charge ratio which coincides with ion mass only for singly charged ions. When normal mass spectra are plotted, or an instrument has the MIKE configuration, the mass-to-charge ratio of all ions being collected can be obtained in the conventional way. For the high-voltage scanning technique, the mass-to-charge ratio of the collected ions is not scanned but in the case of mass analysis of an IKE peak, it is necessary to calibrate the mass scale. In order to see how this is effected when the voltage across the electric sector is changed, consider the following general case. A stable ion of mass m carrying a single positive charge has a kinetic energy

$$eV = \frac{1}{2}mv^2$$

after falling through the accelerating field. As shown in eqn. (3), it will follow a radius in the magnet depending on its momentum-to-charge ratio and this is given by

$$\frac{mv}{e} = \left(\frac{2mV}{e} \right)^{\frac{1}{2}} \tag{20}$$

This momentum-to-charge ratio locates the mass-to-charge ratio m/e on the conventional mass scale. If we now consider a transition in which an ion of mass m_1 carrying x positive charges dissociates into an ion of mass m_2 with y positive charges, the electric sector voltage will need to be changed from E to

$$E_1 = \frac{m_2 x}{m_1 y} \, E \tag{21}$$

in order to transmit the daughter ions. The kinetic energy of the daughter ions is given by

$$\frac{xm_2}{m_1} \, eV = \frac{1}{2} \, m_2 v_2^2$$

and their momentum-to-charge ratio is

$$\frac{m_2 v_2}{ye} = \left(\frac{2Vm_2^2 x}{em_1 y^2} \right)^{1/2}$$ (22)

From eqns. (20) and (22) it can be seen that the daughter ions of actual mass-to-charge ratio m_2/ye are detected by the magnet with an apparent mass of

$$m = \frac{m_2^2 x}{m_1 y^2} = \frac{m_2}{y} \frac{m_2 x}{m_1 y} = \frac{m_2}{y} \frac{E_1}{E}$$ (23)

Equation (23) shows that the readings of mass-to-charge ratio indicated by, say, a mass-marking device operated by the magnetic field need to be corrected by a factor equal to the ratio of the actual electric sector voltage to that appropriate for transmission of stable ions.

If no automatic mass-marking facility is available, it is necessary to refer back to the normal mass spectrum in order to establish the mass scale. In this case, the magnetic field is scanned until the peak due to daughter m_2^+ ions is seen. As soon as this occurs, and without interrupting the magnet scan, the electric sector is switched back to its normal value E. Peaks in the normal mass spectrum are then scanned until the mass scale can be recognized and the apparent mass of the daughter ions determined. This technique is illustrated in Fig. 19 which shows a mass-analyzed IKE peak from n-decane obtained with an electric sector voltage of 200.0 V. The mass scale with the normal electric sector voltage of 347.4 V gives an apparent mass of 32.8 from which the true mass of the ions is seen to be 57.0 after multiplication by the correction factor for the change in electric sector voltage. Mass analysis may be carried out on a double-focusing mass spectrometer of Mattauch—Herzog geometry in which the spectrum is recorded on a photographic plate. A normal spectrum and an image due to daughter m_2^+ ions formed when either the accelerating voltage or the electric sector voltage is changed can be used to give successive exposures on the plate. The mass can be read directly from the plate. Conversion to the true mass value is necessary only in the case where the electric sector voltage has been changed. Figure 20 shows the result of mass analysis of overlapping IKE peaks centered at $1.355E$ in the spectrum of biphenyl. By setting the accelerating voltage to 73.8% (1/1.355) of its normal value, the ions which com-

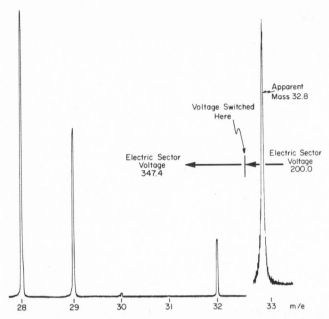

Fig. 19. Mass analysis of an ion kinetic energy peak. The ions giving rise to the IKE peak to be analyzed are focused at the β-slit and then allowed to pass through the magnet by raising the β-multiplier. The mass spectrum is scanned and, after the daughter ion is recorded, the electric sector voltage is switched to the main ion beam where the mass spectrum is recorded without interruption of the mass scan. The actual mass of the daughter ion can be obtained from the apparent mass (m/e 32.8) by multiplying by the ratio of the electric sector voltages.

Fig. 20. Mass analysis of an IKE peak using a Mattauch—Herzog instrument. Three exposures are reproduced; (a) and (c) show a portion of the normal mass spectrum of biphenyl, and (b) is that obtained by reducing the accelerating voltage to 73.8% of its normal value. The mass-to-charge ratio of the lines due to the metastable ions can be read from the figure. The true masses involved can be read directly because the electric sector voltage has not been changed.

prise the IKE peak will be transmitted with E set to its normal value. Mass analysis of this ion beam shows that three ions of mass-to-charge ratios 69.5, 87 and 102 contribute to this peak. Since $m_1/x = (m_2/y)(V_1/V)$, the first transition is characterized as

$$\frac{m_1}{x} = \frac{69.5}{1.355} = 51.3 = \frac{154}{3}$$

thus

$$154^{3+} \rightarrow 139^{2+} + 15^{+}$$

A variety of other methods of characterizing the ions involved in metastable transitions has been used and we shall mention two that pertain to double-focusing sector instruments in which the electric sector precedes the magnetic sector. In the first of these, due to Major[173], it is necessary that sufficient ions decompose in front of the magnetic sector that a metastable peak can be detected in the normal mass spectrum. The magnet current is set to collect this peak and then the electric sector voltage is changed until another peak due to daughter ions of the metastable transition occurring in the first field-free region is seen. The magnet is, throughout, accepting ions of apparent mass $m* = m_2^2/m_1$. The first peak is due to ions of mass m_2 formed as daughters in the second field-free region. When the electric sector is suitably adjusted to a new value $E_1 = (m_2/m_1)E$, daughters m_2^+ formed in the first field-free region and having the correct momentum to be recorded as being of mass $m*$, will be transmitted. Thus, we have that $m_2 = m_1 E_1/E$ and $m_2^2/m_1 = m*$ from which m_2 (and hence m_1) can be determined uniquely.

In the second scanning method, the MIKE scan is simulated without exchanging the position of the electric and magnetic sectors. For simplicity, the method is illustrated for singly charged ions; it can readily be generalized to accommodate multiple charges. A peak due to ions of mass m_1 is collected at the final collector after the magnetic sector with the accelerating voltage and sector voltage set to their normal values. We have already seen that it is possible to transmit daughter ions of mass m_2 formed in the first field-free region either by increasing the accelerating voltage or decreasing the electric sector voltage. More generally, the daughter ions will pass through the electric sector if the ratio of accelerating to electric sector voltage is changed from its normal value K to a new value $K(m_1/m_2)$. However, in general, the momentum of the daughter ions will depend on the value of the accelerating voltage; if the accelerating voltage is changed from V to V_a where $V_a/V = m_1^2/m_2^2$, the momentum of the daughter ions will be given by eqn. (22) as

$$m_2 v_2 = \left(2V_a \frac{m_2{}^2}{m_1}\right)^{1/2} = (2m_1 V)^{1/2}$$

But this is the momentum of the metastable parent ions of mass m_1, so the $m_2{}^+$ ions can be collected by the magnet. But while the accelerating voltage is changed from V to V_a, the sector voltage must be changed from E to E_a where $E_a/E = m_1/m_2 = (V_a/V)^{1/2}$. Thus $E^2/V = $ constant. Only by scanning in this way will the ratio V/E change to the value $K(m_1/m_2)$ necessary for the daughter ions to pass through the electric sector. Combined scanning of V and E in this way can, therefore, be used to bring all the daughter ions of $m_1{}^+$ successively to the final collector, thus simulating the MIKE scan.

ADVANTAGES AND DISADVANTAGES OF THE VARIOUS SCANNING METHODS

In considering the advantages of scanning in different ways we shall compare, in each case, an IKE scan without mass analysis, an HV scan with subsequent mass analysis and a MIKE scan.

We have already seen that overlapping of peaks can often occur in an IKE spectrum. This overlapping is due to various values of parent and daughter masses giving identical or closely similar ratios. In the HV scan method, a single species is selected from the several daughter ion masses that can pass the electric sector and so this difficulty is largely avoided. There may still be a problem in a very small number of cases if two daughter ions of the same nominal mass but different empirical formula are formed or if two identical daughter ion species can be produced by two different fragmentation processes. In the MIKES case, too, this difficulty is largely circumvented; in this case it is only ions from a parent (or group of parents of equal nominal mass but different formulae) whose reactions are studied at one time.

The IKE scan gives an excellent method of "fingerprinting" a particular organic compound. It is generally true that a greater number of diffuse peaks are produced in a mass spectrometer than normal mass peaks; thus the IKE spectrum gives a plot from which a large amount of structural information can be extracted and by which slight differences in structure can be detected as discussed in Chapter 5. From the IKE spectrum, the important regions which require further study are quickly apparent. Thus, an IKE scan can provide background information that can be used in HV scan or MIKE studies. The latter two techniques cannot give the overall "fingerprint" except by a computer-aided synthesis following a comprehensive set of scans at individual mass-to-charge ratios.

A disadvantage of the HV scan method compared with the other two is due to changes produced in the field within the ion source when the accelerating voltage is changed. This "detuning" of the source leads to difficulties when relative abundances of daughter ions have to be determined or when low-abundance ions have to be studied. To overcome this difficulty, it is necessary to tune the instrument for each individual peak. That is to say, the value of accelerating voltage, V_1, corresponding to the peak is determined. The electric sector voltage is adjusted so that the main beam of stable ions is transmitted at voltage V_1 and tuning is carried out. The electric sector voltage is then returned to its normal value, without retuning, and the HV can then be scanned over a small range at maximum sensitivity. The difficulty does not arise in IKE or MIKE studies where the accelerating voltage is not scanned.

A second disadvantage of scanning the accelerating voltage is due to the limited range over which such scans can be carried out. Most double-focusing mass spectrometers can be operated without difficulty over a ratio of accelerating voltages of 4:1 and with special precautions over a ratio of 8:1. This is adequate for studying the majority of fragmentation reactions; in most cases, less than half the mass of the parent ion is lost as the neutral fragment but, from time to time, cases do arise where it is required to examine a very low-mass daughter ion and in such cases the HV scan method fails. There is no such limitation for the other two methods where only the electric sector voltage is changed; this voltage can be scanned to zero if necessary.

An important consideration that can affect the relative sensitivity attained with the three methods is concerned with the absence of z-axis focusing in most commercial instruments. The electrostatic and magnetic sectors act only as cylindrical lenses and the beam spreads linearly in the z-direction as it moves through the mass spectrometer tube (see Fig. 5). In the IKE technique, the beam is detected closer to the ion source than is the case in the other two methods. Thus, for a fixed slit height, the IKE method is the most sensitive. It will be seen later in this chapter that slits cannot be made of indefinite height; the attainable resolution is affected by curvature of the image produced by the sector fields and, for metastable ions that decompose with conversion of internal energy into kinetic energy, the attainable energy resolution is an inverse function of the slit height. In cases where sensitivity is of prime importance, however, the IKE technique gives best sensitivity. The MIKE technique can be more sensitive than the HV scan method because in the latter case the beam is severely restricted in height by the small gap in the magnetic field. When this field is positioned in front of the final collector, it influences the effective slit height that can be used whereas, in the MIKE arrangement, the electric sector, which is the final sector traversed by the beam, does not restrict slit height in this way.

It has been pointed out above that if the unimolecular decomposition of a metastable ion in a field-free region takes place without conversion of internal energy into kinetic energy of separation of the fragments, then the metastable peak that is observed will be distinguishable from a peak due to the arrival of stable ions at the collector only by virtue of its appearing, in the general case, at a non-integral mass number. Before considering the effect of energy release during fragmentation on metastable peak shape, we shall consider how fragmentations of metastable ions within regions of the flight path in which there is an electric or magnetic field may also contribute to the observed mass spectrum.

FRAGMENTATION IN NON-FIELD-FREE REGIONS

Let us consider, throughout, the case of an ion of mass m_1 carrying x positive charges that decomposes in a unimolecular reaction into a daughter ion of mass m_2 carrying y positive charges. Let us first consider a fragmentation occurring during acceleration of the ions emerging from the ion source. If the parent ion has fallen through a fraction cV of the main accelerating voltage, V, before decomposing, then the daughter ions will be formed with a kinetic energy $xceV(m_2/m_1)$. They will then be accelerated through the remainder of the accelerating field and emerge with a kinetic energy given by

$$\tfrac{1}{2} m_2 v^2 = xceV \frac{m_2}{m_1} + y(1-c)eV$$

In a similar manner to that used in deriving eqn. (16), above, for a single-focusing mass spectrometer, we may write for the apparent mass-to-charge ratio of the $m_2{}^+$ ion that is transmitted through the magnetic field

$$\frac{m^*}{e} = \frac{1}{e}\frac{m_2^2}{m_1}\left(\frac{m_2xc + m_1(1-c)y}{m_2y^2}\right) \tag{24}$$

An equation of this general form was first developed by Hipple *et al.*[141]. From eqn. (24), it can be shown that if the ions of mass m_1 decompose at a potential very close to that of the ion source ($c \cong 0$), then the mass-to-charge ratio of the daughter ions is given by $m^*/e \cong m_2/ye$. These ions will appear in the low-mass tail of the daughter ion peak. If the parent ions fall through almost the whole of the accelerating potential before decomposing ($c \cong 1$) then the apparent mass-to-charge ratio is $m^*/e \cong (m_2^2/m_1)/(x/y^2)$. These ions appear in the high-mass tail of the metastable peak due to decompositions in the field-free region. Decompositions at intermediate values of c will give rise to a continuum of ions that fall between these two extreme apparent mass values. The actual distribution of ions across the continuum will depend upon the shape of the accelerating field and the half-life of the decomposition, although it can be expected that because the ions leave the source essentially with zero velocity, a significant percentage of the decompositions will occur near $c = 0$. It is difficult to observe the continuum between $(m_2^2x)/(m_1y^2)$ and m_2/y because it is overlaid by another continuum due to decompositions that occur within the magnetic sector and which is discussed below (p. 56). In a double-focusing instrument, if adjustments are made to the electric sector voltage to enable the range of kinetic energies due to decompositions in the accelerating field to be studied, a peak is also seen due to decompositions that occur within the electric sector. (This is discussed later in this chapter.)

It has already been explained that ions that decompose in the field-free region in front of the electric sector will, in general, lose suffcient kinetic energy that they will fail to pass through the energy-resolving slit (the β-slit) that follows this sector. Ions that decompose within the electric sector itself will also suffer discrimination, but if they decompose near the exit end of the electric sector they will be deflected less and may be transmitted through the β-slit. In the MS9 mass spectrometer manufactured by AEI, for example, an ion kinetic energy range of $\pm0.66\%$ is transmitted through the β-slit. In the RMH-2 mass spectrometer manufactured by Hitachi, the width of the β-slit is variable, the maximum spread of ion kinetic energies that can be transmitted being $\pm0.2\%$.

It is instructive to examine a mass spectrum as the electric sector or accelerating voltage is changed progressively and the result of doing this is illustrated in Figs. 21—23, the β-slit width being adjusted so as to transmit an energy range of ±0.1%. Figure 21 shows the result of varying the accelerating voltage when toluene is used as the sample. The spectrum (a) of the figure shows a magnet scan over

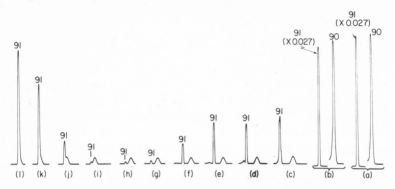

Fig. 21. Effect of varying the accelerating voltage on the position of peaks in the mass spectrum of toluene. The diffuse peak for the transition $92^+ \rightarrow 91^+$ is due to fragmentations within the electric sector and it can be seen to move progressively into the peak at mass-to-charge ratio 91 which is due to decompositions within the accelerating field. The sensitivity of the scans is × 1 unless otherwise labeled.

masses 90 and 91 with normal accelerating voltage (8 kV) to sector voltage ratio. It can be seen that the peak at mass 91 is very large (it is the base peak of the spectrum) and that a metastable peak can just be distinguished under the normal peak at mass 90. The accelerating voltage is then increased progressively in steps of about 0.1%, a partial mass spectrum being plotted at each accelerating voltage value. The next plot (b) is very similar to the first, the β-slit being just wide enough to accommodate a change of 0.1% in the energy of the ions issuing from the electric sector. Plot (c) is, however, very different. The peak at the position of the normal mass peak (mass-to-charge ratio 91) is reduced in size by a factor of about 74; the normal peak at mass-to-charge ratio 90 is no longer detectable and only the metastable peak due to loss of mass 1 from mass 92 ions can be seen at this position. As further scans are carried out [(d)—(j)] the narrow peak at mass 91 decreases by a further factor of about 20 and then begins to increase again. The metastable peak changes comparatively little in height or shape but its position on the mass scale moves until it begins to merge with mass 91. The peak at mass 91 increases in size by a factor of about 50 between scans (i) and (l) where the accelerating voltage reaches 92/91 of its normal value. At this value of accelerating voltage, ions of mass 92 that decompose into ions of mass 91 anywhere along the entire length of the first field-free region will be given sufficient extra energy to be transmitted through the electric sector. They will be transmitted through the magnet as ions of mass 91 with normal momentum and will thus appear at mass-to-charge ratio 91. It has been seen above that ions that fragment in the accelerating field will lose less energy than those that decompose in the first field-free region after acquiring the full accelerating energy. Ions that fragment early in the accelerating field will lose less energy than those that decompose later. As the accelerating field is increased in the series of scans (a)—(l), the daughter ions of fragmentations in the accelerating region are transmitted, those that decompose earliest being transmitted in the earlier scans. These ions pass through the magnet like normal ions of mass-to-charge ratio 91. They are well-focused and give rise to a narrow peak because they all appear to have passed through the source slit and thus will focus on the β-slit.

Suppose, now, that some ions fragment at the exit of the electric sector. These daughter ions will be of mass 91 but carry only a fraction 91/92 of the normal ion energy. They will thus give rise to a metastable peak of apparent mass $91^2/92 \cong 90$. An ion that decomposes near the exit of the electric sector will traverse most of the sector with full energy. To correct its path for the small distance it would travel with reduced energy, the accelerating voltage has only to be increased slightly. Furthermore, the resultant small increase in final kinetic energy above the fraction 91/92 of the normal main beam energy brought about by the increase in accelerating voltage will change the apparent mass very slightly above 90. Looked at in the other way, as the accelerating voltage is increased, daughter ions formed earlier and earlier within the electric sector will be transmitted and their apparent mass will slowly increase. The limit is reached when ions formed in front of the electric sector are detected at an accelerating voltage 92/91 of normal; these ions are recorded as being of mass 91. Thus the narrow and diffuse peaks can be used, respectively, to give kinetic information concerning fragmentations along the length of the accelerating field and along the length of the electric sector, respectively. The total number of scans is too small to show complete details of all the variations, but it is interesting to see that there is not a smooth fall in the height of the narrow peak between scans (c) and (i), scan (e) being anomolously large. This is due to there being a field-free region (in the form of a lens) part way along the acceleration path in the RMH-2. Thus a larger number of ions have an energy associated with fragmentation at the potential within this lens. A small peak that appears superimposed on mass 91 in scan (c) and that moves upwards in mass whilst decreasing in size in scans (d) and (e) is an artifact due to reflection of a small number of the ions that make up the main beam from the electric sector wall without neutralization of their charge. If the intermediate (IKE) detector were to have been used, a much larger peak due to neutrals reflected from the walls would have been seen[47].

To illustrate further the occurrence of fragmentations within the

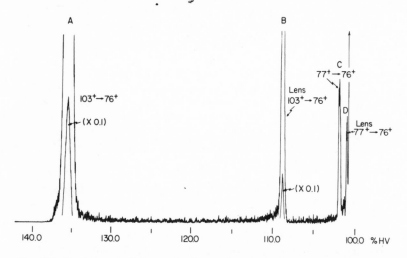

Fig. 22. Abundance of the peak observed with the magnet set to transmit ions of mass-to-charge ratio 76 as the accelerating voltage is progressively increased in the spectrum of benzonitrile. The reactions $103^+ \rightarrow 76^+$ and $77^+ \rightarrow 76^+$ occurring both in the first field-free region and in the lens (third field-free region) can be seen.

lens, consider Fig. 22 which shows a plot of the abundance of the peak at m/e 76 in the spectrum of benzonitrile as the HV is progressively increased. The largest peak (A) occurs at a voltage 103/76 of the normal value and corresponds to the fragmentation of molecular ions in the first field-free region. Peak B is due to fragmentations of molecular ions within the lens. Peaks C and D are due to fragmentation of ions of m/e 77 to m/e 76 in the first field-free region and the lens, respectively.

Figure 23 shows the series of curves corresponding to those for toluene shown in Fig. 21 as obtained when the accelerating voltage is

sector itself[38, 84, 215]. It can be shown that, for a sector magnetic field, the dissociations result in a continuum which may extend from the mass of the parent ion m_1 to the apparent mass of the metastable ion m^*. The intensity of the continuum rises slightly at each limit. However, if the masses of the parent and daughter ions differ by more than a few units, daughter ions formed near the center of the sector may strike the walls and be lost. For a typical commercial instrument and a small mass loss ($\sim$ 1%), the complete continuum may be observed. Figure 24 shows the continuum in toluene for the transition $92^+ \rightarrow 91^+$. The continuum due to fragmentations within the magnetic sector covers the mass range 92 to 90 and is clearly visible. Comparison of the shapes of the lower mass sides of parent and daughter masses sometimes shows a distortion at the low mass side of the daughter mass[38]. This is due to daughter ions formed

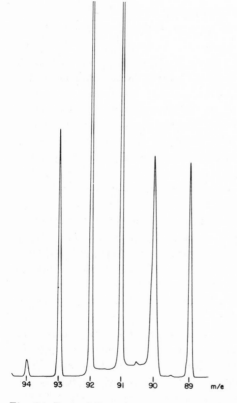

Fig. 24. Part of the mass spectrum of toluene showing the continuum between mass-to-charge ratios 90 and 92 due to the transition $92^+ \rightarrow 91^+$ occurring within the magnetic sector.

within the accelerating field near enough to the ion chamber to be transmitted through the electric sector although they have slightly less than the energy of normal daughter ions.

Finally, it may be noted that fragmentations occurring immediately in front of the final collector without release of internal energy cannot be detected by the shape or position of the peak. If internal energy is converted into kinetic energy, however, it has been shown that this can give rise to "tailing" of the observed peak at the mass of the parent ion.

THE EFFECT OF ENERGY RELEASE ON PEAK SHAPE

All the foregoing discussion of the focusing of the daughter ions of metastable transitions has assumed no conversion of internal (excitation) energy into external (kinetic) energy and the conclusion has been reached that such reactions occurring in the field-free regions will lead to metastable peaks as sharply focused as normal mass peaks. A study of any mass spectrum of an organic compound such as that shown in part in Fig. 15 will show, however, that metastable peaks are invariably diffuse and can be of various shapes. The diffuse nature of the peaks is due entirely to conversion of internal energy into kinetic energy of separation of the fragments. The narrowest metastable peak so far observed[54] releases less than 2×10^{-4} eV of energy, ($\sim$ 4 cal/mole), and is perceptibly broadened compared to the normal mass peaks by this energy release (see Fig. 26, p. 61). The detailed shapes of metastable peaks are of interest in a wide range of mass spectrometric studies, but comparison of practically observed peaks with those predicted by mathematical calculation is still in its infancy due to the extreme difficulties in the calculation caused by the range of speeds and directions acquired by the daughter ions, by the consequent variety of paths described through the focusing fields, by the fact that decompositions occur throughout the length of the field-free region and in the fringing fields, by the finite slit lengths and widths over which integration of ion abundances must be carried out, by the range of energies released and by the range of half-lives involved. The apparently hopeless situation can, however, be dealt with to a certain extent and the general features of the observed peaks can be understood by some simple mathematics. It is convenient, in order to determine the velocity of the product ion in its most general form, to utilize a coordinate system moving with the center of mass of the parent ion m_1^+ (that has a velocity v_1 along the x direction). Let u_2 and u_3 be the velocities of the daughter ion

$m_2{}^+$ and the neutral fragment m_3 relative to the center of mass and let the direction of u_2 be defined by the angles θ and ϕ as shown in Fig. 25. If T measures the internal energy of $m_1{}^+$ that is converted into

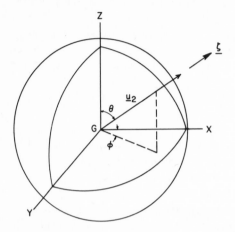

Fig. 25. Direction of motion of fragment ions $m_2{}^+$ defined by the angles θ and ϕ in the center of mass system. The velocity of the ions is given as u_2.

translational energy in the fragmentation, then by the law of conservation of energy

$$T + \tfrac{1}{2}\, m_1 v_1{}^2 = \tfrac{1}{2}\, m_2\, (v_1 + u_2)^2 + \tfrac{1}{2}\, m_3\, (v_1 + u_3)^2$$

Expanding, this reduces to

$$T = v_1 \cdot (m_2 u_2 + m_3 u_3) + \tfrac{1}{2}\, m_2 u_2{}^2 + \tfrac{1}{2}\, m_3 u_3{}^2 \qquad (25)$$

By the law of conservation of momentum, relative to the center of mass we have that

$$m_2 u_2 + m_3 u_3 = 0 \qquad (26)$$

Thus, eqn. (25) reduces to

$$T = \tfrac{1}{2}\, m_2 u_2{}^2 + \tfrac{1}{2}\, m_3 u_3{}^2 = \frac{m_1 m_2}{2m_3}\, u_2{}^2$$

Or,

$$u_{2\cdot} = \left(\frac{2m_3 T}{m_1 m_2} \right)^{\!1/2} \varsigma$$

where ζ is a unit vector along the direction $\mathbf{u}_2$. Converting to a fixed coordinate system, the velocity of m_2^+ is given by

$$\mathbf{v}_2 = \mathbf{v}_1 + \mathbf{u}_2$$

The resulting component velocities of $\mathbf{v}_2$ in the x, y and z directions can thus be written

$$v_{2x} = v_1 + \left(\frac{2m_3 T}{m_1 m_2}\right)^{1/2} \sin\theta \cos\phi \tag{27}$$

$$v_{2y} = \left(\frac{2m_3 T}{m_1 m_2}\right)^{1/2} \sin\theta \sin\phi \tag{28}$$

$$v_{2z} = \left(\frac{2m_3 T}{m_1 m_2}\right)^{1/2} \cos\theta \tag{29}$$

Thus, the daughter ion m_2^+ is given additional components of velocity along the three coordinate axes. The maximum and minimum velocities possible in the original direction of motion of the ions m_1^+ are given from eqn.(27) as

$$v_{2x(\max)} = v_1 + \left(\frac{2m_3 T}{m_1 m_2}\right)^{1/2} \quad \text{and} \quad v_{2x(\min)} = v_1 - \left(\frac{2m_3 T}{m_1 m_2}\right)^{1/2}$$

It can be seen that the ions continuing to move exactly along the original direction of motion with a range of velocities represented by $v_1 \pm (2m_3 T/m_1 m_2)^{1/2}$ and thus with corresponding spreads of momenta and of kinetic energy will be transmitted through a magnetic sector or an electric sector over a range of field strengths. If we assume that it is the ions having this extreme range of velocities along the x direction that form the extremities of the resultant mass or IKE peak, then the extreme width of the peak may be related to the energy release, T, in the following manner[51]. The kinetic energy of these m_2^+ ions is given by

$$\tfrac{1}{2} m_2 v_2^2 = \tfrac{1}{2} m_2 \left[v_1 \pm \left(\frac{2m_3 T}{m_1 m_2}\right)^{1/2} \right]^2 = \tfrac{1}{2} m_2 v_1^2 \left[1 \pm \left(\frac{2m_3 T}{m_1 m_2 v_1^2}\right)^{1/2} \right]^2$$

$$= \tfrac{1}{2} m_1 v_1^2 \frac{m_2}{m_1} \left[1 \pm \left(\frac{m_3 T}{m_2 xeV}\right)^{1/2} \right]^2$$

$$= xeV\frac{m_2}{m_1} \left[1 + \frac{m_3 T}{m_2 xeV} \pm 2 \left(\frac{m_3 T}{m_2 xeV}\right)^{1/2} \right] \tag{30}$$

The radius of curvature of this daughter ion in a magnetic sector is given by

$$r = \frac{m_2 v_2}{Bey}$$

Thus

$$\frac{B^2 r^2 e}{2V} = m^* = \frac{x m_2^2}{y^2 m_1} \left[1 \pm \left(\frac{m_3 T}{m_2 xeV} \right)^{1/2} \right]^2$$

$$= \frac{x m_2^2}{y^2 m_1} \left[1 + \frac{m_3 T}{m_2 xeV} \pm 2 \left(\frac{m_3 T}{m_2 xeV} \right)^{1/2} \right] \qquad (31)$$

For typical accelerating voltages of several thousand volts and a typical energy release of the order of 1 eV, $m_3 T/(m_2 xeV)$ may be neglected compared with $\{m_3 T/(m_2 xeV)\}^{1/2}$. If $m_3 = m_2$, $T = 1$ eV and $xeV = 10^4$ eV, the term $m_3 T/(m_2 xeV)$ will shift the apparent mass position of the center of the metastable peak by an amount $10^{-4} (m_2^2/m_1) = 5 \times 10^{-5} m_1$. This is generally too small to be detectable. The shift becomes smaller if m_2 and m_3 are not equal. If d is the extreme spread of the metastable peak in mass units, from eqn. (31) we can write

$$d = 4 \frac{x m_2^2}{y^2 m_1} \left(\frac{m_3 T}{m_2 xeV} \right)^{1/2}$$

or

$$T = \frac{y^4 m_1^2 d^2 eV}{16 x m_2^3 m_3} \qquad (32)$$

The maximum fractional spread of energy of the m_2^+ daughter ions can also be written as $\Delta E/E_1$ where ΔE is the range of values of electric sector voltage over which the peak is observed in an IKE or MIKE scan and E_1 is the sector voltage corresponding to the center of the peak. Now, $E_1 = Exm_2/(ym_1)$, so

$$\frac{\Delta E}{E_1} = 4 \left(\frac{m_3 T}{m_2 xeV} \right)^{1/2} = \frac{\Delta E}{E} \frac{m_1 y}{m_2 x}$$

Thus

$$T = \frac{x m_2 eV}{16 m_3} \left(\frac{\Delta E}{E_1} \right)^2 = \frac{y^2 m_1^2 eV}{16 x m_2 m_3} \left(\frac{\Delta E}{E} \right)^2 \qquad (33)$$

As an example of the way in which this equation is used, consider the IKE peak from *sym*-triazine shown in Fig. 26. There are two superimposed peaks and we shall estimate the energy release from the narrow one. This is due to unimolecular decomposition of the molecular ion. The wide peak arises from collision-induced decom-

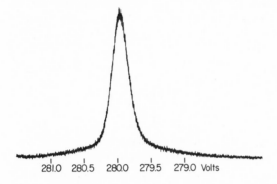

Fig. 26. Ion kinetic energy peak due to the transition $81^+ \to 54^+$ in *sym*-triazine.

positions which are discussed later in this chapter. After making a small correction for the contribution to the peak height from the wide component, the width of the peak is estimated to be 0.273 V at half-height. The electric sector voltage corresponding to the peak center, E_1, can be seen to be 280 V. The electric sector voltage corresponding to transmission of the main beam, E, is 420 V and the corresponding accelerating voltage V is 5250 V. The transition is due to loss of neutral HCN from molecular ions in the reaction

$$81^+ \to 54^+ + 27$$

so that $x = y = 1$, $m_1 = 81$, $m_2 = 54$ and $m_3 = 27$. If there were no energy release in the transition one would expect the energy in the daughter ion beam to show the same percentage spread as the main beam itself. Since the energy of the daughter ion beam is exactly 2/3 that of the main beam, its width at half-height without energy release would be 2/3 of the main beam width (0.186 V) *i.e.* 0.124 V. This voltage is thus subtracted from the observed width in volts of the daughter ion beam at half height, giving a value relating to the energy release of 0.149 V (ΔE). Inserting these values into eqn. (33)

$$T = \frac{81^2 \times 5250}{16 \times 54 \times 27} \left(\frac{0.149}{420}\right)^2$$

$$= 1.86 \times 10^{-4} \text{ eV}$$

If the alternative method of HV scans is used, T can similarly be calculated. If V_1 is the accelerating voltage necessary to give daughter ions of energy yeV, then the extreme values of V to transmit the entire peak will be given by

$$\frac{\Delta V}{V_1} = \frac{1}{1-2\left(\dfrac{m_3 T}{m_2 xeV}\right)^{1/2}} - \frac{1}{1+2\left(\dfrac{m_3 T}{m_2 xeV}\right)^{1/2}}$$

$$= 4 \left(\frac{m_3 T}{m_2 xeV}\right)^{1/2} + 16 \left(\frac{m_3 T}{m_2 xeV}\right)^{3/2} + \dots$$

To a close approximation,

$$\left(\frac{\Delta V}{V_1}\right)^2 \cong \left(\frac{\Delta E}{E_1}\right)^2$$

and

$$V_1 = \frac{m_1 y}{m_2 x} V$$

Thus

$$T \cong \frac{xm_2 eV_1}{16m_3}\left(\frac{\Delta V}{V_1}\right)^2 = \frac{x^2 m_2{}^2 eV}{16ym_1 m_3}\left(\frac{\Delta V}{V}\right)^2 \tag{34}$$

Equations (32)—(34) thus enable T to be estimated in normal mass spectra, IKE or MIKE spectra and HV scans, respectively.

As an example of the calculation of the kinetic energy release using the HV scan technique, consider the reaction

$$C_6 H_6{}^{+\cdot} \rightarrow C_6 H_5{}^{+} + H^{\cdot}$$

in benzene. It is experimentally convenient to measure a fixed fraction of the ion accelerating voltage rather than the full value. All the voltages given below, therefore, refer to a fraction (0.02634) of the full HV and should be multiplied by 37.966 to convert to actual voltages. On this scale, the initial value of the accelerating voltage was 180.438 V and the value at which the center of the metastable peak was transmitted was 182.781 V. The metastable peak width at half maximum was 0.280 V. This has to be corrected for the main beam width measured at the same accelerating voltage (182.781 V) as is necessary to transmit the metastable fragmentation product. The correction is adjusted for the fact that only a fraction, m_2/m_1, of the energy spread in the main beam can be transmitted to the daughter ion m_2^+. The main beam width at half-height was 0.025 V at 182.781 V which is barely changed (to 0.0247 V) on multiplication by m_2/m_1. Hence, the corrected metastable peak width is 0.280–0.0247 = 0.2553 V. Substituting in eqn. (34)

$$T = \frac{77^2 \times eV}{16 \times 78 \times 1} \left(\frac{0.2553}{180.438} \right)^2$$

Now, the kinetic energy of the stable ion beam in electron volts is 180.438 x 37.966. Hence, in electron volts,

$$T = \frac{77^2 \times 180.438 \times 37.966}{16 \times 78 \times 1} \left(\frac{0.2553}{180.438} \right)^2$$

$$= 65 \times 10^{-3} \text{ eV, } i.e. \text{ 65 meV}$$

An alternative version of eqn. (34) which is valid for an instrument in which the fraction 0.02634 (1/37.966) of the accelerating voltage is monitored, is therefore

$$T = \frac{x^2 m_2^2 \; 37.966 \; (\Delta V)^2}{16 \; y \; m_1 \; m_3 \; V}$$

where ΔV and V are expressed in terms of this fraction of the accelerating voltage.

It is instructive to consider the range of T that can be studied with a typical high-performance instrument such as the RMH-2. The main beam of ions, when scanned across the β-slit, shows a width at half-height of the order ± 0.5 eV for an accelerating voltage of 10 kV. This width is due to a combination of energy spread in the beam and imperfections in focusing. It is possible, with care, to detect a change in the width at half-height equal to about 5% of this width, $i.e.$ equivalent to ± 0.03 eV. The spread of kinetic energies of the m_2^+ ions is given by eqn. (30) as $(4xeVm_2/m_1)\{m_3T/(m_2xeV)\}^{\frac{1}{2}}$. The internal energy released by the m_1^+ ions leading to this spread is T eV. We can thus define an "amplification" factor, A, determined by the ratio of these two quantities as

$$A = \frac{4xeVm_2}{Tm_1} \left(\frac{m_3 T}{m_2 xeV} \right)^{\frac{1}{2}} = 4 \left(\frac{xeV}{T} \right)^{\frac{1}{2}} \frac{(m_2 m_3)^{\frac{1}{2}}}{m_1} \tag{35}$$

This amplification factor depends upon the half-power of the ratio of the kinetic energy acquired during acceleration and the internal energy converted, multiplied by a factor depending upon the mass lost in the fragmentation. The form of the function $(m_2 m_3)^{\frac{1}{2}}/m_1$ as m_2 and m_3 vary is shown in Fig. 27. The maximum amplification possible when 10 keV ions $m_1{}^+$ are used is, for example, 100 when $T = 1$ eV, 1000 when $T = 10$ meV and 10,000 when $T = 100$ μeV. The fact that the amplification factor increases as T becomes smaller is an advantage when the limit of detection is approached. If T were as

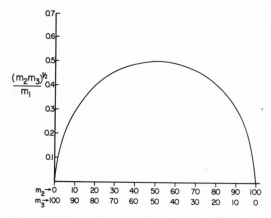

Fig. 27. Form of the function $(m_2 m_3)^{\frac{1}{2}}/m_1$ where m_1 is the mass of the parent ion and m_2 and m_3 those of the fragments formed from it.

small as 10^{-6} eV, the amplification factor of 10^5 would still mean that, in the best case, the $m_2{}^+$ ions would have a range of kinetic energies from 4999.9 to 5000.1 eV. In the absence of any energy release, it would be expected that the percentage spread in energy of the $m_2{}^+$ daughter ions would be the same as that for the $m_1{}^+$ parents. One would thus expect a spread of ± 0.25 eV for these 5 keV ions at half-height. The range of kinetic energies produced when $T = 10^{-6}$ eV would thus give a 40% increase in the width of the peak at half-height and would easily be detectable. A release of 10^{-6} eV of internal energy is far less than would be expected in any fragmentation. Thus, all metastable peaks can confidently be expected to be broader than normal mass peaks and this broadening can readily be detected in any high performance instrument.

The various metastable peak widths shown in Fig. 15 are accompanied by a variety of peak shapes[36, 38, 112, 136, 137, 214]. As systems are examined in which the release of energy is greater than for the processes shown in Fig. 15, metastable peaks develop "flat" tops and for even greater energy releases such as shown in Fig. 28, the peak tops can become "dished". The reason for these effects can be found by examining eqns. (27)—(29). The behavior of those ions for which $\theta = \pi/2$ or $3\pi/2$ and $\phi = 0$ or π (see Fig. 25) has already been discussed. They will acquire no velocity components in either

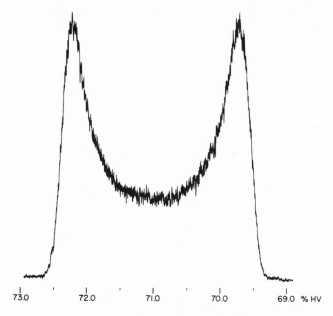

Fig. 28. Metastable peak associated with a large release of kinetic energy. The peak shown was plotted by the HV scan technique.

the y or z directions. Other ions will correspond to $\theta = \pi/2$ or $3\pi/2$ and $\phi = \pi/2$ or $3\pi/2$ and acquire maximum deflection in the $x–y$ plane while ions for which $\theta = 0$ or π and $\phi = \pi/2$ or $3\pi/2$ will receive maximum deflection in the $x–z$ plane. The majority of the ions will, of course, correspond to intermediate values of θ and ϕ. A deflection in the $x–y$ plane, corresponding to the acquisition of a y-component of velocity will cause an increase in the divergence of the beam, but provided this does not lead to the beam width becoming broader than any limiting slit (the α-slit), variation of the accelerating voltage or electric sector voltage will bring the ions through the β-slit. The quantitative effect of this y-deflection, which causes the ions to follow a non-central path through the radial electric field of the electric sector or the field of the magnetic sector, is not easy to calculate, but it will certainly contribute to the peak shape. Ions that receive a deflection in the $x-z$ plane receive a component of velocity in the direction parallel to the lengths of the slits (the z-direction). Most mass spectrometers have no focusing action in this direction and if the deflection is large enough, the ions can fail to pass the relevant collector slit. The maximum deflection in the z-direction is received by ions that have received no velocity components in the x or y directions during fragmentation. If they do succeed in passing through the collector slit they will, therefore, be recorded at the center of the metastable peak. Conversely, discrimination against deflected ions, especially likely in mass spectrometers with short slits, will be greatest at the center of a metastable peak and this accounts for the flattened or dish-topped appearance often observed. The discrimination will increase as T increases; as explained above and illustrated in Fig. 29, it will become more pronounced as slit lengths are reduced; it will also become greater as the distance between field-free region and collector is increased. If peak shapes

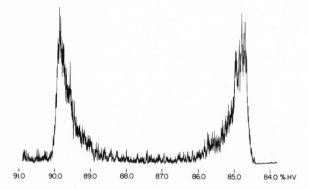

Fig. 29. Shape of a peak obtained by the HV scan technique using short slits. The peak corresponds to the reaction $91^{2+} \rightarrow 52^+ + 39^+$ in toluene. The discrimination in the z direction is so great that no ions are collected in the center of the peak.

obtained by the IKE technique are compared with those from HV scans with mass analysis, the latter will give more sharply dished peaks, merely because the beam has further to travel before collection. Discrimination will also increase as the maximum angle of deflection of the ions increases due to the released energy becoming a greater percentage of the accelerating energy. These points are illustrated in Figs. 30 and 31. It can also be seen from Fig. 31 that as the accelerating voltage is reduced, the width of the peak changes inversely with the half power of this voltage. Reducing the accelerating voltage so as to broaden the peaks on the mass or energy scale does not make measurement of the amount of energy released any more accurate; as the width of the peak increases, its shape changes, the

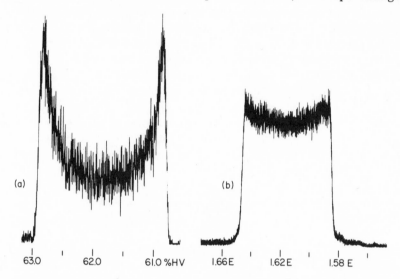

Fig. 30. Shape of the peak due to the transition $78^{2+} \rightarrow 63^+ + 15^+$ in benzene plotted by (a) the HV scan technique and (b) the IKE technique. It can be seen that the discrimination at the center of the peak is greater in (a) because of the longer distance traveled by the ions before collection.

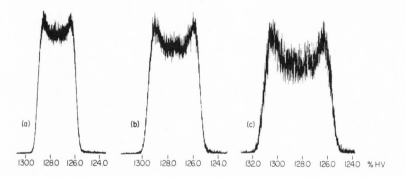

Fig. 31. Shape of the peak for NO˙ loss from the molecular ion of o-nitrotoluene obtained by the HV scan method. As the accelerating voltage is progressively reduced, from (a) 5.9 kV to (b) 3.4 kV and (c) 2.0 kV, the width of the peak increases and it becomes more "dished" in appearance.

slope of the sides becoming less steep. We have already discussed the problems that arise due to the overlapping of peaks in IKE spectra. It is an advantage in most cases to work at the highest possible electric sector voltage (and, hence, accelerating voltage). This will narrow the individual peaks and lead to the highest energy resolution and least overlapping of peaks. An example of the improvement in resolution that can be achieved in this way is shown in Fig. 32. This shows part of the IKE spectrum of naphthalene in the region of $0.5E$. It can be seen that the two narrow IKE peaks become separated from the more diffuse IKE peak as E is increased progressively.

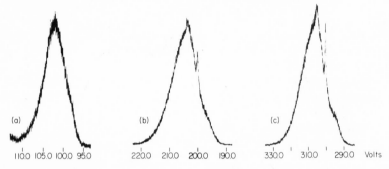

Fig. 32. Progressive improvement in resolution obtained for a peak collected at the β-multiplier as conditions are adjusted so that the main beam of ions is transmitted at higher voltages.

The above discussion of the broadening of metastable peaks due to the release of internal energy during fragmentation has been greatly simplified. The detailed calculation of peak shapes by calculating the paths of individual daughter ions that have been displaced from the central path through the electric or magnetic sector by energy release has only recently been attempted[58]. Analytical expressions for the ion flux falling on a collector slit of particular dimensions cannot be found unless gross simplifications are made to the mathematical equations, but the use of digital computers is now making it possible to synthesize peak shapes to be expected from various instruments. Examples of some calculated IKE peak shapes are shown in Fig. 33. These all relate to hypothetical reactions in which every fragmentation releases exactly the same amount of energy, T. In a practical case, some deviation from this ideal situation must always be expected; there will be a probability distribution governing the relationship between the range of rate constants appropriate to decomposition within the field-free region and the non-fixed (excitational) energy, $\epsilon^{\ddagger}$. The fraction of the total excess energy available that is actually

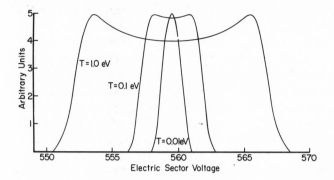

Fig. 33. Metastable peak shapes calculated for the transition $139^+ \rightarrow 109^+ + 30$ with the release of 1.0, 0.1 and 0.01 eV of kinetic energy. The calculations were carried out for the geometry of the RMH-2 mass spectrometer.

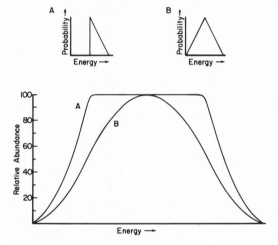

Fig. 34. Effect of the hypothetical probability functions A and B on the shape of a rectangular metastable peak.

released in the fragmentation is itself dependent on several factors that are discussed in Chapter 4. Even if the shape of the metastable peak were rectangular for release of a single amount of energy, complicated changes in shape could be produced by various probability functions for energy release. Two of these are illustrated in Fig. 34. In the first, the probability of a particular minimum energy release is taken as unity and the probability is assumed to fall linearly to zero at twice this energy value. In the second, the probability is assumed to rise linearly from zero to a maximum value and then to fall back to zero at the same rate. The second distribution results in a more sharply peaked metastable. It can readily be shown that if the observed distribution for a particular energy release were dish-topped, then a probability function such as B in Fig. 34 would result in some filling-in of the dish as well as producing a lowered slope on the sides of the peak. It is often difficult, in a practical case, to resolve differences in metastable peak shapes even when these are caused by the superposition of two peaks due to processes releasing quite different amounts of energy. An example is given in Fig. 35 which shows filling-in of a dished peak top by a second peak due to a process releasing less energy. The best energy resolution is always obtained with the shortest possible slits and with collimation of the beam to within a narrow angle. This leads to a peak shape for a single energy release approximating to two separate spikes, a situation being approached in the example already shown in Fig. 29. Unfortunately, as discussed below, these are conditions which result in low sensitivity, limiting the range of transitions that can be observed.

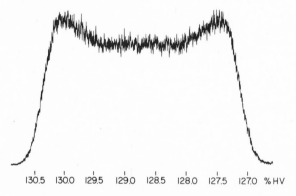

130.5	130.0	129.5	129.0	128.5	128.0	127.5	127.0	% HV

Fig. 35. Shape of the peak due to the transition $M^{+\cdot} \rightarrow (M-NO)^{+}$ in p-nitrophenol plotted by the HV scan method. The broad dish-topped peak can be seen to be superimposed on a second narrower peak that fills in the center of the dish although no details of this narrower peak can be discerned.

REQUIREMENTS FOR OBSERVING METASTABLE PEAKS AT HIGH SENSITIVITY OR HIGH RESOLUTION

Metastable peaks are invariably broader than normal mass peaks; IKE peaks are invariably broader than the peak due to the main beam of normal ions. Under these circumstances, the height of the metastable peak (or the IKE peak or the HV scan peak) may be increased relative to the peaks from stable ions by widening the

collector slit. This can be seen by reference to Fig. 36 which shows in
(a) two hypothetical metastable peaks and two normal mass peaks
produced by scanning with a collector slit of the width shown. The

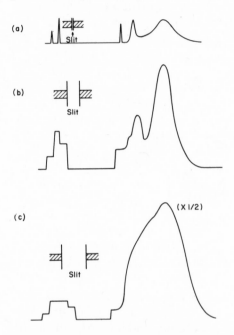

Fig. 36. Effect of widening the collector slit upon the shape of normal mass
peaks and metastable peaks. It can be seen that the relative heights of metastable
peaks of different widths vary with slit width. The effect on resolution of wide—
ning the slits can be seen both for normal and for metastable peaks.

mass peaks are shown as being well-resolved, while the metastable
peaks are just separated from one another. As the collector slit width
is increased, the normal mass peaks develop flat tops and an inverted
valley between them; but the heights of the flat tops of the individual peaks and of the inverted valley do not increase. The metastable
peaks begin to merge, but their heights increase relative to the normal mass peaks. Thus, in order to observe metastable peaks at high
sensitivity, the current must be integrated over as great an area of the
peak as possible by widening the collector slit. For this reason, few
metastable peaks are seen in high-resolution mass spectra. A compromise between sensitivity and resolution is, however, always necessary. As can be seen from Fig. 36, as the metastable peaks overlap
more and more, it becomes increasingly difficult to locate them. The
interference between metastable peaks and normal mass peaks will

also tend to increase as the collector slit is opened; it will, therefore, become more difficult to detect any small metastable peaks that lie in the skirts of the peaks due to stable ions.

The improvement in sensitivity of detection of metastables that can be obtained by opening the collector slit is greatest for broad metastables, *i.e.* those that are formed with the greatest release of kinetic energy and are therefore, most likely to overlap with other peaks. To reduce discriminations in the z-direction in these peaks, the collector slit may also be lengthened, but this has the effect of filling in the center of a dish-topped peak and reducing the amount of structural detail that can be observed. Let us consider the effect of an α-slit, placed in the first field-free region, upon the sensitivity with which a decomposition in this region may be detected. For observation of the maximum number of daughter ions from fragmentations that release an appreciable amount of kinetic energy and, therefore give ions that may be deflected from their original direction of motion, this α-slit should be wide. But this will allow off-axis ions that are deflected back nearer to the central direction of motion to be collected. This, too, will destroy useful structure in the metastable peak shape. For high sensitivity of detection, the field-free region should be long so that a large number of fragmentations can occur in it. But fragmentations occurring near to the front face of the sector will not focus correctly at the β-slit and will also fill in the center of a broad peak. Thus, high sensitivity of detection and high resolution in the sense of avoiding overlapping of peaks or of picking out fine details of peak structure, are mutually exclusive.

There are, of course, several basic points to be borne in mind. (*i*) The greater the number of ions leaving the ion chamber within a narrow range of kinetic energies, the greater will be the signal-to-noise ratio for a particular process. Under these circumstances, collimation of the ion beam can be improved if desired. (*ii*) The more sensitive the detector and the longer the time taken over the measurements, whether by use of long time constants and slow scans, by signal averaging or ion counting, the more accurate the results will be. (*iii*) The larger the radius of electric and magnetic field employed, the larger the slits can be made without too much degradation of performance. (*iv*) A double-focusing instrument will give higher sensitivity than a single-focusing instrument of similar size because a greater kinetic energy spread can be included in the main beam and brought to a good focus at the final collector.

A related problem arises when one attempts to measure the energy released in a metastable transition. A correction must be made for the width of the main beam of ions. One simple method is to assume (as was done on p. 61) that in the absence of any energy release, the percentage spread in energy will be the same in the main beam and in the peak obtained in an IKE or HV scan. For the IKE scans, therefore, the main beam width is measured and multiplied by E_1/E, where E_1 corresponds to the center of the peak being studied. For HV scans, the main beam width is measured at the accelerating voltage, V_1, necessary to transmit the product of metastable ion fragmentation, and this width is multiplied by a factor equal to V/V_1 (*i.e.* m_2/m_1) to adjust the energy spread in the main beam for that transmitted to the product ion m_2^+. The resultant value is applied as a correction by subtracting it from peak width. If the main beam width is small compared with the IKE peak, the error in this procedure is negligible; but when the energy release in the IKE peak is of the order of 1 meV or less, it can become considerable. It is preferable, of course, to deconvolute the main beam shape out of the observed peak-shape, but this cannot be done if one does not even know the peak shape to be expected when the same amount of energy is released each time an ion fragments. One is calculating the energy from the width measured at half-height merely as a convenient compromise. A measurement of this kind will, however, become more meaningful if it is used merely to compare energy releases in similar systems or to confirm an identical pattern of energy release in two transitions.

One further factor that might cause important variations in peak shape is concerned with the profile of the ion flux density arriving at the plane of the β-slit. Figure 37 shows several idealized cases. In (a)—(c), a uniform flux density is assumed to arrive at the β-slit. Curve (b) shows that if the β-slit width is exactly equal to the width of the image, a triangular peak will result when the distribution is scanned across the slit. Parts (c) and (a) show that a flat top develops in the peak if the β-slit is either narrower or wider than the ion flux distribution. In practice, it is usual to narrow the β-slit progressively until any flat top disappears and signal strength begins to be lost, *i.e.* condition (b) is reached. If however, the ion flux density is not uniform, it may be more difficult to set the width of the β-slit. Parts (d)—(f) show the situation where the β-slit is, respectively, wider, equal to and narrower than the ion flux when the shape of the flux density is a triangular function; it can be seen that the change in shape in moving from (e) to (f) is less easy to detect. The situation for a gaussion-shaped ion flux density is shown in (g)—(j). The transition from (h) to (j) is not easy to follow. Part (k) of Fig. 37 shows a

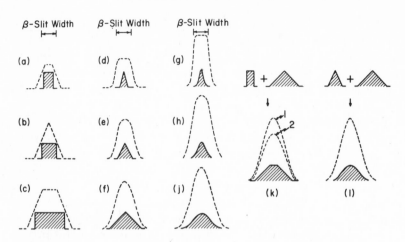

Fig. 37. Effect of different ion flux densities scanned across β-slits of the same width. The flux densities are shown shaded and the resultant peak shapes are drawn as dotted lines. Only the widths are drawn to scale; relative heights are in arbitrary units.

rectangular main beam ion flux distribution convoluted with a wider, triangular, IKE peak shape. The dotted curve (1) shows the peak shape when the β-slit is slightly wider than optimum while curve (2) shows the situation when the β-slit width is the same as that of the main beam. Part (1) shows a similar convoluted ion density for a triangular main beam density and IKE peak of twice main beam width. From the resultant shape of the IKE peak obtained for the "correct" β-slit width [as in (e)] one can see that there is no easy, intuitive way of correcting the signal for the effect of the main beam. Reproducible results are obtained, however, if the HV or IKE peak is plotted at several β-slit widths and the peak width at half-height is extrapolated to zero slit width. The extrapolation procedure is also carried out for the main beam of stable ions and the metastable peak width is corrected in the usual way (see above). This rather lengthy procedure is only necessary when the difference between main beam and metastable peak widths is small. This is usually the case for reactions in which H· is lost, since the amplification factor is then small, or when the energy release is unusually small, as for the loss of HCN from *sym*-triazine.

Special techniques may be used in some instances when especially high sensitivity or especially high resolution is required. The number of ions drawn from the source may be increased by the methods used in chemical ionization[109, 213]; the source is made essentially gas tight and differential pumping added so that sample pressures can be increased. Ions can be drawn from the ionizing region by viscous flow and very high sensitivities are achieved in this way. Alternatively, a very large number of ions can be produced (currents of the order of mA) by the use of a duo-plasmatron source. The ion current leaving the source is then so large that it can be collimated to the extent that vibrational fine structure has been observed in the metastable peak (Fig. 38) formed by fragmentation of molecular ions of NO^+ into $O^{+\cdot}$ (ref. 119). Vibrational structure has also been reported[87] in the case of the fragmentation of H_3^+ into H^+ which is discussed further in Chapter 4. The neutral moiety in this reaction is a molecule of hydrogen and this can be formed in any of the vibrationally excited states of the electronic ground state. Maximum

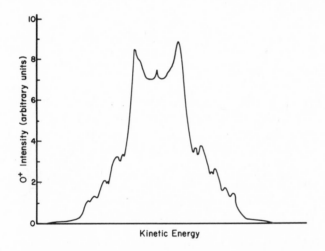

Fig. 38. Appearance of fine structure in a peak due to the transition $NO^+ \rightarrow O^{+\cdot} + N^{\cdot}$. The structure is due to reactions of NO^+ ions in various vibrational states.

kinetic energy is released when the molecule is formed without vibrational energy; the probability of forming the product with various amounts of vibrational energy is reflected in the shape of the HV scan peak which is composed of overlapping peaks of various widths.

Structure may also be seen in IKE peaks for a completely different reason. When the same daughter ion is formed from two different parents of slightly different masses, the resulting IKE peaks may differ only slightly in position. For example, in the case of methane in the presence of a collision gas, (the effect of which is discussed below), an ion of m/e 1 is formed from ions of m/e 16, 15, 14 and 13. This gives rise to almost coincident peaks centered at $E/16$, $E/15$, $E/14$

and $E/13$, respectively, all of different heights. The shape of the combined peak is shown in Fig. 39 and the position of the overlapping maxima is indicated.

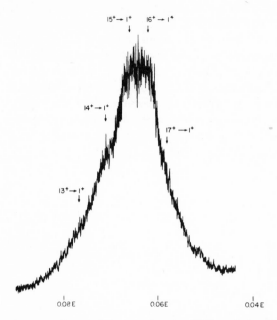

Fig. 39. Shape of the overlapping IKE peaks in the spectrum of methane due to the transitions $16^+ \rightarrow 1^+$, $15^+ \rightarrow 1^+$, $14^+ \rightarrow 1^+$, and $13^+ \rightarrow 1^+$. A small peak corresponding to $17^+ \rightarrow 1^+$ and due to fragmentation of $CH_5{}^+$ ions can just be discerned.

It has already been mentioned that a detector that can be used to measure a range of mass-to-charge ratios or ion energies simultaneously gives a greatly increased sensitivity over the same detector used to measure only a narrow range at one time and with which the spectrum must be scanned. Thus, the photographic plate (even though it requires several hundred ions to produce a perceptible image) can compete with the much more sensitive electron multiplier detector (which can respond to a single ion) because it can record simultaneously over a wide mass range. Recent development of arrays of miniature electron multipliers such as the spiraltron strip detector developed by Bendix or the two-stage chevron detector, enable individual signals to be picked up from high-gain multipliers arranged in an array in which the individual members are only 3 μm apart. A scanning procedure and time averaging of the detected signals enables an array of several hundred signals to be examined at the highest sensitivity. The electric sector on, say, the RMH-2 mass spectrometer gives a dispersion of 5 μm at the β-slit for an energy change of $10^{-3\,5}$ so that, bearing in mind the "amplification" factor discussed above, this is adequate for studying any practical system. In fact, the energy resolution of which such an electric sector is capable in practice is an order of magnitude less than this.

Even without special sources and detectors, high sensitivity of detection of metastable peaks can be achieved. A range of IKE peak heights from 2×10^4 : 1 can easily be achieved as shown in Fig. 40 which illustrates the sensitivity obtained in an HV scan on benzonitrile. Peak B, for example, is due to the loss of mass 27 from molecular ions of benzonitrile of mass 106, that is molecular ions which contain three isotopic labels at natural abundance.

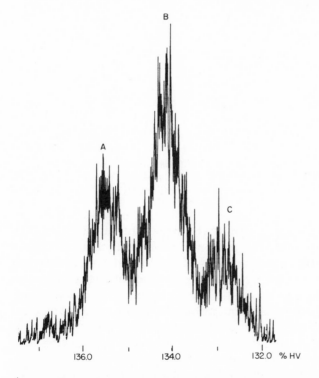

Fig. 40. Fragmentation of benzonitrile molecular ions containing several isotopic labels at natural abundance plotted by the HV scan technique. Peak A corresponds to loss of mass 28 from ions of mass 107 [a combination of $(M+1)^+$ ions containing 3 heavy isotopic labels and ions $M^{+\cdot}$ containing 4 heavy isotopic labels at natural abundance]. Peak B is due to the reaction $106^+ \rightarrow 79^+$ and peak C to the reaction $105^+ \rightarrow 79^+$.

COLLISION-INDUCED DISSOCIATION

If the background pressure in the mass spectrometer is allowed to rise to a value of the order of 10^{-4} to 10^{-3} Pa($\sim 10^{-6}$ to 10^{-5} torr)[124, 154, 232], the number and the abundance of diffuse peaks observed in the mass spectrum of an organic compound both increase rapidly. These peaks are due to fragmentations induced in stable ions by collision with neutral molecules of background gas and they can be used to give information on molecular structure as discussed in Chapter 5. The nature of the collision gas is generally not important and it is a simple matter to construct a sample introduction system with which collision gas can be introduced at a controlled pressure[53]. Problems of sensitivity and resolution involved in the detection and measurement of these peaks [generally referred to as collision-induced dissociation (CID) peaks] are the same as for metastable peaks. However, CID peaks are not due to the fragmentation of metastable ions; they arise only because sufficient internal energy is given to the ions in the field-free region that they are caused to fragment in a range of times from those as short as for ions that are formed within the ionization chamber to several microseconds. Because collision, and, hence, fragmentation, can take place anywhere in the collision region, useful structural features in the shapes of CID peaks will be lost. The cross-section for the collisions is, however, high, and can often be of the order of 10^{-13} cm^2. There is thus no difficulty in constructing a much shorter collision region and operating it at a higher pressure. Ideally, the collision region needs to be as close as possible to the entrance slit to the field-free region and to be arranged with a system of differential pumping so that collision gas does not escape into the rest of the vacuum system. A typical arrangement that could be used is illustrated schematically in Fig. 41. In this case no sensitivity need be lost by using the shorter region; the collision gas pressure is merely increased by the appropriate amount. In the case of unimolecular decompositions of metastable ions, too, an improved peak shape would result if the field-free region were deliberately shortened so that it was restricted to a portion of the ion path much closer to the entrance slit. This could, in principle, be accomplished as shown in Fig. 42. The new field-free region would be restricted to the small length of the enclosed region between the two slits. The potential of the field-terminating plate directly at the entrance of the electric sector would be raised to as high a value, (X volts), as possible. Thus, as the daughter ion left the short field-free region with an energy

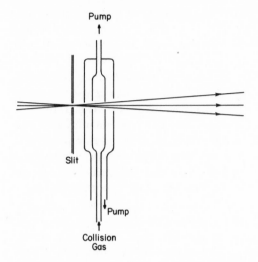

Fig. 41. Schematic diagram of a differentially pumped collision chamber of short length located near the entrance slit to a field-free region.

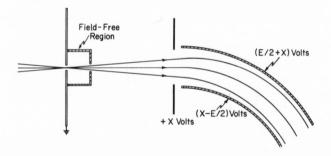

Fig. 42. Schematic diagram of a short field-free chamber located at the front part of the first field-free region. The remainder of the region in front of the electric sector has a superimposed field caused by the potential difference between the slit (ground) and the front of the electric sector.

$(m_2/m_1)\, eV$, it would come under the influence of a repulsive field and would finally enter the electric sector with an energy $e[(m_2/m_1)V-X]$. Ions that decomposed after leaving the field-free region but before entering the electric sector would have a range of kinetic energies from the above value to $em_2(V-X)/m_1$ and would thus be spread over the plane of focus of the electric sector contributing little to the main peak shape. The sensitivity of detection of metastable peaks would, however, be reduced in this method.

The amount of energy released during an ion fragmentation in a field-free region offers a method of distinguishing between unimolecular and collision-induced reactions. If a peak suspected to be from the unimolecular decomposition of a metastable ion shows no change in relative abundance or shape as a collision gas is introduced into the field-free region [except for a slow fall off in abundance beyond a pressure of about 10^{-3} Pa (10^{-5} torr) due to scattering of the ion beam], this is certain evidence that the parent ions of the transition are indeed metastable. On the other hand, if the peak increases linearly in height with the collision gas pressure, this is evidence that a CID is being observed; it provides no evidence, however, concerning the possibility of there being a weak metastable peak under the collision-induced component. To search for such a metastable peak the observed peak height must be plotted as a function of pressure down to the lowest pressure obtainable. Non-linearity is the only way of detecting the metastable peak; cross-sections for collision-induced reactions are sufficiently high that, even at pressures of the order of 10^{-6} Pa (10^{-8} torr), CID's can be observed.

In many cases, both a unimolecular and collision-induced component of a peak are easy to observe. Fig. 43 (a) shows a peak at m/e 28 obtained in an HV scan arising from the fragmentation of doubly charged molecular ions of carbon dioxide. The dish-topped peak ob-

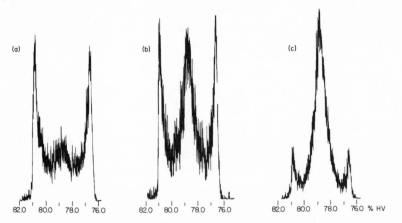

Fig. 43. Peak of mass-to-charge ratio 28 in the spectrum of doubly charged carbon dioxide obtained by the HV scan technique. In (a) the fragmentation $44^{2+} \rightarrow 28^+ + 16^+$ predominates; in (b) the collision-induced process $44^{2+} \rightarrow 44^+ \rightarrow 28^+ + 16$ contributes to the center of the peak; in (c), at higher collision gas pressure, this latter process dominates.

served is, as we have already seen, due to release of a large coulombic energy due to separation of the charges. In (b) a small pressure of nitrogen is introduced as collision gas and a collision-induced peak can be seen near the center of the dish top. As the collision gas pressure is further increased, this collision-induced component predominates. It is worth remarking that even though a peak of m/e 28 is being observed, the use of nitrogen as collision gas poses no problem. The ratio of the accelerating voltage being used to that required to transmit the main beam of ions is given by

$$\frac{V_1}{V} = \frac{m_1 y}{m_2 x} = \frac{44}{56}$$

and only ions that have 56/44 of the energy of the main beam can be transmitted. Such ions can only come from doubly charged molecular ions by loss of $O^{+\cdot}$.

In this example, the width of the collision-induced peak at half-height is smaller than that of the unimolecular component. This is a situation commonly met with only in the case of multiply charged parent ions. A charge-exchange reaction has preceded fragmentation in the presence of the collision gas so that the large coulombic energy is not available for release in this case. Generally, however, for transitions of ions carrying a single charge, the energy release is greater for the CID peak than for the metastable peak. This is mainly due to the greater range of internal energies that can be given to the ion, while the fragmentation can still be observed, in the CID case. The collision itself takes place in the field-free region and so very fast fragmentations are included among those observed, the only criteria for observation being that the fragmentation should also occur in the field-free region and that the ion should not be scattered out of the beam in the collision. A typical case, in which both the metastable and CID peaks can be seen, is shown in Fig. 44. A composite peak is also apparent in Fig. 26. It should be noted from Fig. 44 that the centers of the two component peaks are not coincident. This and other properties of CID peaks are discussed in more detail in Chapter 4. A cautionary note should be added before ending this section; there are potentially many more collision-induced decompositions than metastable transitions in the majority of mass spectra of organic compounds. Leakage of sample out of the ion chamber into the field-free region is often sufficient to generate large numbers of these ions. Several papers claiming to discuss metastable ions but which actually deal with CID's have already been published.

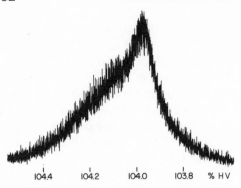

Fig. 44. Unimolecular and collision-induced fragmentation $(26^+ \to 25^+ + 1)$ giving rise to superimposed peaks in the HV scan of mass 25 in acetylene. Note that the two reactions do not give product ions which have exactly the same kinetic energies.

EQUIPMENT FOR IKES AND MIKES WORK

As will be apparent from the preceding paragraphs, any single- or double-focusing sector mass spectrometer offers the opportunity of studying reactions of metastable ions. No commercial equipment specifically designed for this work has yet appeared, although in the last few years, one double-focusing mass spectrometer of "reversed" geometry in which the magnetic sector precedes the electric sector has made its appearance[179] while other manufacturers have offered accessories particularly useful in this kind of work. It may be useful to the reader considering the acquisition of equipment to work in this field to consider the features of the two instruments currently in use in the authors' laboratory, one a modified commercial instrument and the other an instrument designed especially for MIKES work. The first of these instruments is shown in schematic form in Fig. 45 and consists of a modified RMH-2 instrument[44] made by Hitachi—Perkin Elmer. The instrument is of modified Nier—Johnson geometry. It is fitted with a direct inlet probe and heated all-glass inlet system, has a magnetic sector of 400 mm radius and an electric sector of 500 mm radius. Both sector angles are 70°. The ion source is of the "bright ion" type and does not include a source magnet. The maximum recommended source voltage is 9.6 kV. The ion beam issuing from the source is focused onto the source slit by a system of electrostatic lenses. The lens component immediately in front of the ion source consists of two parallel plates each 37 mm long. The space within this lens constitutes the so-called "third" field-free region in the instrument and it can be used to study reactions of metastable

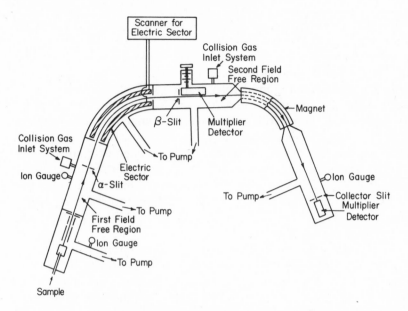

Fig. 45. Ion Kinetic Energy (IKE) Spectrometer. This instrument is a modified double-focusing mass spectrometer of Nier—Johnson geometry.

ions. The instrument wiring has been modified so that the potential of this lens can be varied from its normal value of about 0.7 times the source potential to cover a range from a potential within about 25 V of that of the source to less than 0.5 times the source voltage. The first and second field-free regions are each about 600 mm long, giving very high sensitivity for the observation of metastable ions. The result of being able to follow the decay of metastable ions in three different field-free regions, to alter the time of observation in these regions by varying the lens and accelerating voltages and the ability also to observe fragmentations occurring within the electric sector, makes the instrument a versatile tool for studies in ion kinetics. The source and collector slits are variable in width up to 250 μm; α- and β-slits which are also fitted are variable to ten times this width. All slits can be translated in the y-direction. Originally, α- and β-monitor electrodes were fitted to measure the current passing through the respective regions, but both of these have now been removed. The β-monitor has been replaced by a 10-stage electron multiplier that can be lowered into the beam during IKE scans but raised to allow the beam to pass into the magnetic sector for mass analysis when desired. As well as the pumps for the sample handling system, separate diffusion pumps are used to evacuate the source and lens region, the first field-free region, the electric sector, the second field-free region and the region of the final collector. Each pump is fitted with a valve system by means of which its pumping action can be cut off. Separate sample handling systems are used to introduce collision gas, whenever required, into either the first or second field-free regions via the entrance originally used for the α-monitor and by means of a small hole drilled through the analyzer wall, respectively[53]. The electric sector and accelerating voltages are uncoupled and each can be set to any value within their range or scanned over their entire range or any part of it. The maximum scanning speed available will scan either voltage from zero to its maximum value (or in the reverse direction) in a few seconds. Minimum scanning speed has been set as 1 part in 10^7/sec. The voltage on either plate of the electric sector, the accelerating voltage or the magnet current, can be set to any value with the aid of a six-place differential voltmeter. Slow and medium speed scans can also be monitored using this equipment. In order to maintain as good a vacuum as possible, external heaters are fitted to the source, field-free and sector regions. Background pressures are $\sim 3 \times 10^{-8}$ torr ($\sim 4 \times 10^{-6}$ Pa).

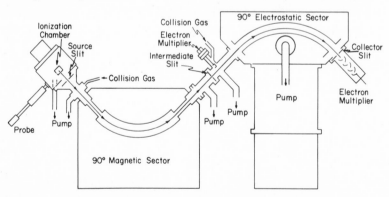

Fig. 46. Mass-Analysed Ion Kinetic Energy (MIKE) Spectrometer.

The MIKE spectrometer, illustrated in Fig. 46, was built after three years' experience of using the RMH-2 and many of the features included in the design are the result of using the latter instrument. The source, magnet and electric sectors are standard parts normally fitted on the MS9 mass spectrometer made by GEC/AEI[93]. The parts have not been positioned in the manner necessary for a double-focusing arrangement, although provision has been made for reversing the deflection through the electric sector relative to that in the magnet, should this prove necessary in later work. High mass

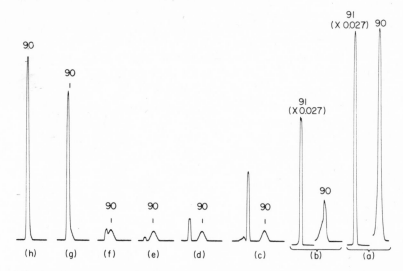

Fig. 23. Effect of varying the electric sector voltage on the position of peaks in the mass spectrum of toluene. The diffuse peaks due to fragmentations within the electric sector appear at mass-to-charge ratio 90, irrespective of electric sector voltage. The sharp peaks due to ions of mass 91 formed in the accelerating region move down the mass scale until they merge with the peak at mass 90.

kept constant, but the voltage across the electric sector is successively reduced by steps of about 0.16%. The only major difference in this case is concerned with the mass scale. As the electric sector voltage is reduced, the sector will pass ions formed earlier and earlier within the sector until finally it will transmit those ions formed in the first field-free region which have a kinetic energy only 91/92 of that of the normal ions. These mass 91 ions of reduced energy will be transmitted by the magnet with an apparent mass 90. This is also the apparent mass of ions formed in the second field-free region. Thus, the peak at mass 91 will move, as the electric sector voltage is reduced, until it coincides with the position of the metastable peak at mass 90. It should be noted that the final peak height recorded [scan (h)] is exactly equal to that obtained when the high voltage was scanned (compare Fig. 21).

If a much wider β-slit were used, the broad peak could cover the whole of the available mass range between mass 90 and mass 91 leading to an apparent continuum between these masses. The effect of scanning the accelerating voltage under these circumstances has been discussed previously.

In addition to decompositions that occur in the second field-free region between the electric and magnetic sectors, and which have already been discussed, ions may dissociate within the magnetic resolution is not vital for MIKES work but high energy resolution is achieved by the present arrangement, the convenience and accessibility of which are extremely important factors. The magnet is of 12 in. radius, the electric sector of 15 in. radius and both sector angles are 90°. The source is of the Nier type, incorporating a magnet, and a maximum accelerating voltage of 8 kV is used. The source and intermediate slit are fixed in width at 5×10^{-3} in. $(1.3 \times 10^{-1}$ mm), the final collector slit is variable from 10^{-4} to 3×10^{-3} in. $(2.5 \times 10^{-3}$ to 7.5×10^{-2} mm). The heights of the intermediate and final slits are adjustable. Two electron multipiers are employed for detecting the beam either at the intermediate or final slits as in the RMH-2. The scanning and monitoring arrangements for the sector and accelerating voltages and for the magnet current are also similar to those developed earlier. Sample introduction via a heated probe or all-glass heated inlet is conventional; collision gas can be introduced into the region between the intermediate slit and electric sector (still called the first field-free region). Three separate pumps are used for evacuating the first and second field-free regions and the region between the magnetic sector and the intermediate slit. An oil diffusion pump is used to evacuate the source. Careful attention is paid to the provision of differential pumping between the evacuated regions, and the arrangement of isolating vacuum valves is the same as described above. Comprehensive baking is provided.

It can be seen that most of the important features of both designs are the same; this emphasizes that there are many factors to be taken into account besides the order in which the beam traverses the electric and magnetic fields.

METASTABLE PEAKS IN OTHER INSTRUMENTS

No reference has been made to the shapes of metastable peaks that can be observed in other types of mass spectrometers such as the crossed-field cycloidal instrument[28]. This omission has been deliberate because we are primarily concerned in this book with applications of metastable peaks to problems in thermochemistry, chemical kinetics, structural and analytical chemistry rather than with a phenomenological approach to the subject. For the same reason, no discussion is included of studies that have been carried out with linear time-of-flight instruments[125]. However, studies using field-ionization sources have provided a wealth of information that complements the knowledge gained from sector instruments fitted with electron bombardment or photon sources, so a short discussion of the experimental technique will be given here. Further discussion is included in Chapter 4. We have already seen that ions that decompose during acceleration can be detected in conventional sector instruments. The unique advantages of the field-ionization source may be used to extend the time range over which observation of reactions in an accelerating field may be made, especially towards much shorter times[23, 98, 150] The basic features of a field-ionization source are shown in schematic form in Fig. 47. A voltage of the order of 5—10 kV is applied to a treated blade, tip or very fine wire placed a few millimeters away from a slit in a metal plate. The sharp edge or

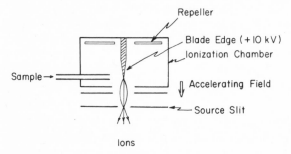

Fig. 47. Field ionization source.

point is at a positive potential with respect to the slit, which forms the entrance slit of a mass spectrometer. More complicated arrangements in which extra electrodes are used to produce a measure of focusing of the beam are often used in practice, but it is unnecessary to discuss these in order to understand the advantages of the method. The effect of the very small radius of curvature is to concentrate the electric field in the vicinity of this small radius as shown in Fig. 48. Fields as high as 5×10^8 V/cm can be produced in the vicinity of the edge and ionization can result. As a molecule approaches the tip and enters the very high field, its Coulomb potential is deformed to the extent that an electron can reach the tip by a process of quantum mechanical tunneling[149]. This process results in the formation of molecular ions containing only very small amounts of internal energy, but fragment ions can still be observed and are thought to be formed by three mechanisms. As well as unimolecular fragmentations of the kind observed in instruments with electron bombardment ion sources and governed by the statistical considerations of QET, ions may fragment from a chemisorbed state at the surface of the field ion emitter[230] or they may be ruptured by a field-induced or field-

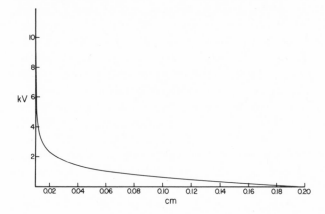

Fig. 48. Plot of potential as a function of distance from emitter tip in a field-ionization source. The field strength at any position is given by the slope of the curve.

assisted dissociation in the very high field region[22]. If all the fragment ions issuing from the source/accelerating region are allowed to enter a magnetic sector for mass analysis, the apparent mass-to-charge ratio of the ions that dissociate in the field will be given by eqn. (24). Ions decomposing from a chemisorbed state for which $c \cong 0$ will appear sharply focused at the value m_2/e, whereas ions decomposed by field-induced or assisted processes will do so over a range of values of c and, hence, will appear spread out on the mass scale over a range m_2/e to $m_2^2/(m_1e)$. Most of the simple cleavage reactions are observed very close to the tip where the field is most intense and where $c \cong 0$. Thus, on its high mass side, the observed fragment peak from a field-induced process will be sharp. On the low mass side, a long tail will be observed corresponding to larger values of c. Because molecular ions are formed where the field is extremely strong, they will be accelerated away from the tip extremely rapidly. That is to say, they will reach a position in the field where c is perceptibly different from zero and, therefore, where they can be observed as a component of the tail, in very short times. An ion of $m/e = 100$ will fall through 10 eV (0.2% of a 5 keV accelerating energy and, thus, detectable) under the influence of a field of 5 V/Å in a time of approximately 4×10^{-14} sec. It is thus possible to study fragmentations from a time as short as this to as much as 2×10^{-7} sec after formation. The ion will then enter the field-free region in front of the magnet, yielding a conventional metastable peak if it fragments there. The shape of the tail observed on the low mass side

of the peak due to the fragment ions of mass m_2 will depend *inter
alia* on the accelerating field and the ion half-life. Just as fragmentations in the lens of the RMH-2 mass spectrometer can be observed
(see p. 54), so in a field-ion source with focusing lenses, intensity
maxima are observed at positions that correlate with the potentials
of the lens components[259]. Beckey *et al.*[23] call these peaks and the
tails themselves "fast metastable peaks". They point out that the
term metastable peak, as generally used, is defined in such a way that
it depends on instrumental parameters in instruments where the ions
take several microseconds to reach the region at which they can be
observed. They are, therefore, justified in adjusting the term to fit
the parameters of field ionization.

One further characteristic of peak shapes in field ionization
deserves special mention. In some cases, when the mass spectrum is
scanned, a small signal is observed at a mass corresponding to m_2,
but the peak is not sharp and rises slowly to a maximum before
tailing down to the apparent mass m^*. This effect of a displaced peak
maximum is characteristic of reactions which occur to their greatest
extent at times corresponding to the molecular ions having fallen
through an appreciable fraction of the accelerating field by the time they
have left the region where the field is most intense. This is the
behavior seen in a number of rearrangement reactions. Some of the
results of measuring the peak shapes in field ionization are discussed
in Chapter 4 where the complementary nature of the information
obtained from electron bombardment and field-ionization studies is
further emphasized.

The properties of metastable ions

"Let me repeat. None of this has any real meaning. . . . We have art in order not to die of the truth"

Camus, *The Myth of Sisyphus*

This chapter deals with the fundamental physical properties of metastable ions and with their reactions. It therefore treats thermochemistry, kinetics and reaction dynamics of unimolecular reactions of gaseous ions occurring within the compass of the mass spectrometer. It also deals with high-energy bimolecular (collision-induced) reactions. Although the ions undergoing such reactions are not metastable, the techniques used and the information gained via collision-induced peaks is similar to that from metastable peaks. Where appropriate, illustrations of the consequences of these properties of metastable ions also have been given, subject to these limitations (*i*) the applications of metastable ions to problems of molecular structure are reserved for Chapter 5; (*ii*) several of the physical properties of metastable ions lead naturally to consideration of the question of ion structure and reaction mechanism. As far as is consistent with clarity, such questions are taken up in the detailed discussion of ion structure which appears in Chapter 6.

APPEARANCE POTENTIAL MEASUREMENTS

For every species of ion in a mass spectrometer formed, say, by electron impact, there is a characteristic threshold electron energy below which the ions do not appear. A plot of ion current for any species against electron energy is known as the ionization efficiency curve for that species. The threshold electron voltage is usually termed the appearance potential of the particular ion. The potential at which molecular ions first appear is called the ionization potential. A typical shape of an ionization efficiency curve in the vicinity of the appearance potential is shown in Fig. 49. The initial curved portion is due mainly to the energy spread of the bombarding electrons. These were produced in the case illustrated by thermionic emission from a rhenium filament at a temperature of about $2000°K$. The electron energy spread is determined by this temperature and also by the potential drop across the filament. The difficulties associated

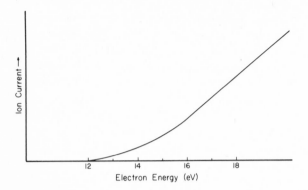

Fig. 49. Ionization efficiency curve in the vicinity of the appearance potential. The curvature at low energy results primarily from the electron energy distribution.

with measuring an appearance potential accurately have been discussed by several authors[82, 161]. These difficulties are partly associated with the problem of detecting the very small currents corresponding to the onset of the processes being studied and partly due to the difficulty of calibrating an absolute voltage scale. The latter difficulty is usually overcome by the use of an internal standard having a known appearance potential. This enables allowance to be made for the voltage drop across the filament, stray contact potentials and field penetration effects into the ionization chamber. A common method of overcoming the first difficulty is to plot ion abundance on a logarithmic scale against electron volts both for the internal standard and for the sample. If the two curves are essentially parallel, the difference in appearance potentials can be determined as the current falls towards zero.

In the production of molecular ions by the process

$$M + e \rightarrow M^{+\cdot} + 2e$$

an electron is expelled from the molecule in a time of the order of 10^{-15} sec. Ionization is said to occur by a "vertical" or Franck–Condon type of process in which inter-nuclear distances remain fixed at the appropriate values for the neutral molecule. The minimum energy necessary to produce molecular ions by electron bombardment is, therefore, called the "vertical" ionization potential and may be greater than the minimum necessary to produce molecular ions in their ground state (the "adiabatic" ionization potential). A fragment ion F^+ may be formed from $M^{+\cdot}$ by the further reaction

$$M^{+\cdot} \rightarrow F^+ + N^\cdot \tag{36}$$

where $N^\cdot$ is a neutral fragment. According to the quasi-equilibrium theory, the appearance potential of F^+ is a measure of the adiabatic ionization potential of M plus the activation energy ϵ_0 that must be given to $M^{+\cdot}$ to produce the ion F^+. The situation is illustrated in Fig. 50. It should be noted that all quantities in the figure are referred to zero-point energies rather than to the equilibrium positions on the potential energy curves.

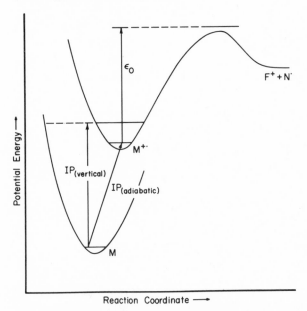

Fig. 50. Thermochemical considerations pertaining to fragmentation of a molecular ion. The activation energy, ϵ_0, is the difference between the zero-point energies of the activated complex and of the molecular ion. The appearance potential of F^+ is the sum of the adiabatic ionization potential and ϵ_0.

In order to observe fragment ions F^+ in the normal mass spectrum, these ions must be formed within the ionization chamber. Typically, an ion spends about 10^{-6} sec in this region before acceleration and, thus, an appreciable number of ions F^+ must be formed in this time interval to ensure their observation. Hence, the observation of F^+ depends not only on the activation energy ϵ_0 being supplied to the molecular ion, but some excess energy sufficient to satisfy the above requirement is also necessary. This excess energy is termed the kinetic shift[82]. It has its origin in the relationship between internal energy and the rate constant for unimolecular fragmentation, but its exact magnitude is dependent upon experimental conditions.

Figure 51 illustrates the relationship between the minimum internal energy (ϵ_{min}) required to cause fragmentation in the ion chamber, the kinetic shift and the activation energies for the forward and reverse reactions. It also shows that for an ion of internal energy ϵ, where $\epsilon > \epsilon_0$, the activated complex will have a non-fixed energy $\epsilon^{\ddagger}$.

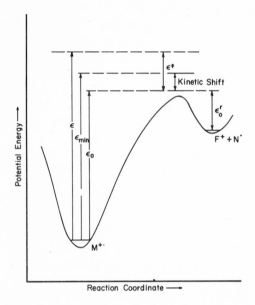

Fig. 51. Definition of some thermochemical quantities pertaining to fragmentation of a molecular ion. The forward and reverse activation energies, ϵ_0 and ϵ_0^r, are the difference between the energy of the activated complex and the ground-state energies of the reactants and products, respectively. If molecular ions $M^{+\cdot}$ have internal energies ϵ exceeding ϵ_0, the activated complex can be attained and will be associated with an internal energy $\epsilon^{\ddagger}$. When $\epsilon^{\ddagger}$ exceeds a minimum energy, termed the kinetic shift, fragmentation occurring in the ion chamber can be observed. The minimum internal energy of $M^{+\cdot}$ required to meet this condition is ϵ_{min}.

It is apparent that the kinetic shift merely represents a particular value of $\epsilon^{\ddagger}$, *viz.* that which is just sufficient to cause detectable fragmentation in the ion source in a given instrument. It should also be noted that in a particular molecule, the Franck—Condon factor may operate so that the ion $M^{+\cdot}$ can only be formed in a high vibrationally excited state such that the minimum attainable value of $\epsilon^{\ddagger}$ is in excess of that dictated by the kinetic shift.

Much of the thermochemical data which has been obtained by

mass spectrometry has relied on two assumptions, (i) the minimum internal energy, ϵ_{min}, necessary to cause fragmentation in the ion chamber, is equal to the activation energy ϵ_0, $viz.$ the kinetic shift can be neglected; (ii) there is no reverse activation energy (ϵ_0^r) for the process. It is one of the important advantages of studies on metastable ions that they allow both these assumptions to be tested and the magnitudes of the excess energy terms ($\epsilon_{min} - \epsilon_0$) and ϵ_0^r to be estimated. Much of this chapter will be concerned with these two problems.

Since the requirements that ions F^+ be formed in the time interval 10^{-6} sec is relaxed if the products of metastable ion decomposition are to be monitored, the appearance potential of the product of a metastable ion should be lower than that of a normal fragment ion. A kinetic shift may still be operative for metastable ion decompositions, but it should be smaller than that for ions which fragment in the ion source. The difference in the appearance potentials provides a lower limit for the kinetic shift operative in ion source reactions.

There is one consideration which might affect the conclusion that the observed appearance potential difference is a measure of the internal energy required to raise the rate of fragmentation from that appropriate to the observation of the products of metastable ion reactions to that necessary to observe normal fragment ions. This concerns the sensitivity with which the measurements are made. Normal daughter ions are detected with high sensitivity and hence it is possible to detect the products resulting from ions fragmenting with rate constant of $\sim 10^4$ sec^{-1}, that is approximately two orders of magnitude smaller than the reciprocal of the average time available for fragmentation. If measurement of the appearance potential of the metastable ion product is done in an instrument of low sensitivity, the minimum rate constant which leads to detectable products may, for example, only be equal to the reciprocal of the average time available for fragmentation, $i.e.$ $\sim 10^5$ sec^{-1}. The high sensitivity available using the HV scan and IKES techniques lowers this rate constant to $\sim 10^3$ sec^{-1}, and, thus, makes it possible to investigate more thoroughly both the existence and the magnitude of the kinetic shift.

Appearance potentials of a number of metastable peaks from organic compounds have been measured at high sensitivity by Hertel and Ottinger[132] and by Hickling and Jennings[135]. The largest difference found between the appearance potential of a normal fragment and a metastable peak was 1.3±0.3 eV in the case of the $C_6H_4^{+\cdot}$ ion

formed from the molecular ion of benzonitrile by loss of neutral
HCN. The kinetic shift appropriate to reactions occurring in the ion
source is, therefore, at least 1.3 eV and has been estimated as 2 eV.
Thermochemical data based on the appearance potential of the
$C_6H_4^{+\cdot}$ ions formed in the source are, therefore, considerably in
error. It must be noted, however, that this represents an extreme
case.

A typical set of ionization efficiency curves for a molecular ion, a
metastable peak and a normal fragment ion is shown in Fig. 52. All
peak heights have been normalized relative to their values at 50 eV.

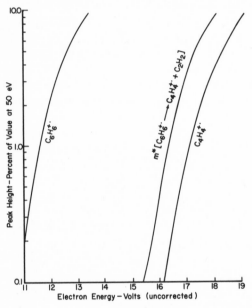

Fig. 52. Ionization efficiency curves for the molecular ion of benzene, the fragment ion $C_4H_4^{+\cdot}$ and the $C_4H_4^{+\cdot}$ ion formed from the metastable molecular
ion.

The curves refer respectively to the molecular ion of benzene, the
normal fragment ion $C_4H_4^{+\cdot}$ in the benzene mass spectrum and the
metastable peak for the reaction

$$C_6H_6^{+\cdot} \to C_4H_4^{+\cdot} + C_2H_2$$

The curves are shown plotted down to an intensity of 10^{-3} of that
at 50 eV. It has been found to be possible to plot metastable peaks
over at least three decades of abundance[46]. It is usually found that

the log abundance *versus* electron energy plots for metastable peaks of a primary fragmentation are closely parallel to those of molecular ions and to that of the inert gas used for calibration purposes over the range 10—0.1% of their abundances at 50 eV. This is so even when the curve for the corresponding normal daughter ion deviates from a parallel curve. In our experience, in every case in which the ionization efficiency curve for the normal daughter ion was of markedly different shape from that for the ion formed in the corresponding reaction of a metastable ion, a reasonable alternative process could always be postulated for formation of the daughter ion. The possible multiple origin of normal daughter ions leads to ambiguities to which studies on metastable ions are not subject. This is true both in thermochemical measurements such as those under consideration, and also in analytical studies, as will be noted in Chapter 5.

At very low abundances ($<$ 0.1% of the abundance at 50 eV) the log abundance plots for metastable peaks sometimes exhibit pronounced curvature and a low-energy tail extending towards the curve for the molecular ion. This tail is attributed to collision-induced dissociation of the molecular ions in the field-free region. The relevant curve for the transition $78^+ \rightarrow 77^+$ in benzene is shown in Fig. 53. The tail can be seen to tend to a straight line of shallow slope which extends down to the ionization potential of the molecular ion. This curve was obtained with an indicated pressure in the field-free region of 1.6×10^{-5} Pa (1.2×10^{-7} torr).

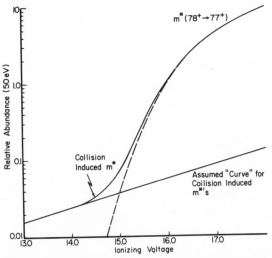

Fig. 53. Ionization efficiency curve at low energies for the ion due to loss of H$^{\cdot}$ from metastable benzene molecular ions. The tail of the curve is due to a collision-induced process.

A further reason for differences in shape between the log abundance plots of molecular and fragment ions is illustrated in Fig. 54 for the case of 1-nitronaphthalene where the curves for ions of mass-to-

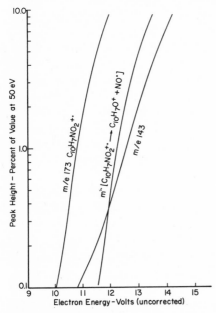

Fig. 54. Anomalous behavior of the ionization efficiency curves for a fragment ion and that for the corresponding metastable peak in 1-nitronaphthalene. The crossing of the curves indicates that m/e 143 is composed of more than one ionic species.

charge ratio 173 (the molecular ion) and 143 are shown, together with the curve for the metastable peak corresponding to the process

$$173^+ \rightarrow 143^+ + 30$$

or

$$C_{10}H_7NO_2^{+\cdot} \rightarrow C_{10}H_7O^+ + NO^{\cdot}$$

At the higher abundances, the curve for the normal fragment ions lies at higher electron energies than the curve for the metastable peak, as would be expected from consideration of a possible kinetic shift. However, at lower abundances, the curves cross over, the curve for the normal fragment ions approaching that for the molecular ions. The crossing of the curves is clear evidence that we are dealing with ions formed by a different process. In this case, ions of mass 143 are being formed by a catalytic reduction of 1-nitronaphthalene within the ionization chamber[46]. The different mass 143 ions are, in fact, molecular ions of α-naphthylamine of formula $C_{10}H_9N^{+\cdot}$ which have a much lower appearance potential than the isobaric fragment ions of formula $C_{10}H_7O^+$ from 1-nitronaphthalene.

Thus, the curves of log ion abundance *versus* bombarding electron energy can be used to study kinetic shifts, to obtain information on collision-induced dissociations and to detect the presence of impurities. It must be pointed out that the ionization efficiency curves of relatively few metastable ions have been measured at a sensitivity sufficient to allow accurate appearance potential determinations from which valid kinetic shift data can be obtained. Consequently, very little has been done in the way of correcting appearance potential data for the kinetic shift.

The foregoing discussion has been limited to a consideration of a single-step fragmentation, implicitly that of lowest activation energy. If two or more processes occur competitively from a molecular ion, then consideration of the rate constants in the range corresponding to fragmentation of metastable ions shows that the largest rate constant will always give rise to the most abundant metastable peak. The process of lowest activation energy will usually, but not necessarily, correspond to this rate constant. Metastable peaks may also be observed for the competitive processes, depending upon the relative magnitudes of the rate constants and the sensitivity with which the measurements are carried out.

KINETICS OF THE FRAGMENTATION OF METASTABLE IONS

Effects of internal energy

Some of the general features of the relationship between k and ion internal energy ϵ will be explored in Appendix I. Particularly important for this discussion of metastable ions is the behavior of k at low internal energies, that is at threshold.

Hertel and Ottinger[132] have considered the relative probabilities that a fragmentation with a given rate constant, k, will lead to the observation of the fragment ion as a normal peak or as a metastable peak. The faster the reaction (the higher the value of k) the more likely it becomes that the fragmentation will occur in the ionization chamber and that the normal peak will be observed. The relative probabilities that a given ion will be detected as a parent ion, a normal daughter ion or as the product of a metastable ion fragmentation, will depend only upon k and on instrumental parameters (compare Fig. 98, p. 230).

A typical pair of curves showing the relationship between k and the intensity of the ion fluxes due to normal fragment ions and to metastable ions, is shown in Fig. 55. At low values of k ($< 10^5$ sec^{-1}), the abundances of the normal fragment ions and the metastable ion products are proportional to log k. In the particular apparatus discussed, the height of the metastable peak was actually greater than that of the normal daughter ion for the transition 103^+ $\rightarrow 76^+ + 27$ in benzonitrile up to a value of k of 3×10^5 sec^{-1}. It

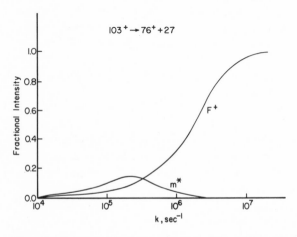

Fig. 55. Relationship between the rate constant, k, for fragmentation and the probability of observing normal fragment ions (F^+) and the corresponding products of metastable ion decomposition (m*) for loss of HCN from benzonitrile.

was found that for values of k greater than 10^8 sec^{-1} there was almost 100% efficiency in detecting normal ions, so that the apparatus could not distinguish k values higher than this. In an actual comparison of the abundances of normal and metastable ions, k can only be varied by changing the excitation energy in the fragmenting ions and this can be accomplished by varying E_{el}, the energy of the bombarding electrons. It is then necessary to make an assumption about how the probability of exciting a specific state, which will react with a rate constant k, varies with E_{el}. Ottinger assumed that the probability varied with the energy ($E_{el} - \epsilon - IP$) possessed by the electrons in excess of that required to reach the state in question. That is to say that the excitation function for each state obeys a linear law near threshold. With the further assumptions that in the vicinity of threshold, log k is proportional to ϵ and that the density of populated states is constant over the log k axis, Hertel and Ottinger[132] were able to compare calculated and experimental intensities and to obtain reasonable agreement. The empirical relationship between k and ϵ which they used to obtain this fit indicates for benzonitrile a suprisingly slow rise of k by only about 1 order of magnitude for an energy interval of 0.65 eV. Figure 55 shows that since the metastable peak is detected over a range of about 10^2 in k, an energy interval as large as 1.3 eV is generating it if the empirical relationship is true. In other reactions for which large metastable peaks are observed, presumably there is also a slow rise of k with ϵ.

Although the above method gave a rough idea of the k *versus* ϵ curve for a particular reaction, it was only recently[4, 5] that an experimental determination of a $k(\epsilon)$ function was achieved. In this work, reactant ions of well-defined internal energy were formed by charge exchange and their fragmentation during acceleration was monitored. Essentially, this kinetic determination is identical in principle to that already described for decompositions in the accelerating region (Chapter 3) and to that used in recent kinetic experiments involving field ionization. The position in the accelerating field at which fragmentation occurs is a direct measure of ion lifetime and is determined from energy analysis of the daughter ion beam. It was assumed that the collection efficiency for fragment ions formed at different positions in the accelerating field was constant. This technique confirmed the early proposal that the rate of increase of k with ϵ for the loss of HCN from benzonitrile was very slow. The appropriate curve is shown in Fig. 56. The validity of the result obtained by this method for $H^{\bullet}$ and C_2H_2 loss from the benzene molecular ion is considered in Chapter 6 in connection with the question of the structure of this ion.

The kinetics of the fragmentation of metastable ions is controlled by the non-fixed energy ($\epsilon^{\ddagger}$) of the appropriate activated complex. Many of the useful features of metastable ions arise because of two properties; (i) metastable ions possess a fairly narrow range of internal energies and (ii) the average internal energy is not much in excess of ϵ_0, that is, $\epsilon^{\ddagger}$ is small[263].

The same limited region at the low energy end of the internal energy distribution of an ion formed, say, by electron impact, is always responsible for metastable peak formation. Hence, the properties of metastable ions allow the characterization of ions in terms of their structure with the minimum influence of internal energy. Specifically, metastable ion abundances (and, as discussed later, kinetic

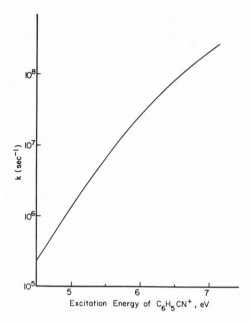

Fig. 56. Experimentally determined relationship between the rate constant and the internal energy, ϵ, of the molecular ion for the loss of HCN from benzonitrile.

energy releases) are largely independent of the overall range of internal energies acquired by the parent ions in the ion source.

The generalization that metastable ion abundances are relatively insensitive to changes in the internal energy distribution, $P(\epsilon)$, in electron-impact experiments has found experimental support[197]. This work, however, also uncovered some cases of pronounced changes in abundance with the relatively small changes in $P(\epsilon)$ produced by varying the ion source temperature. For example, the fragmentation of the 1,2-diphenylethane molecular ion to give $C_7H_7^+$ showed a $[m*]/[M^{+\cdot}]$ abundance ratio of 0.07 at 60°C and 0.89 at 280°. This difference can be explained as the result of the fact that the $P(\epsilon)$ function is not a smooth curve but contains local maxima and minima.

Two methods of expressing metastable ion abundances have found general use. In the first, the metastable ion abundance is expressed relative to the abundance of normal daughter ions (F^+) formed in the same reaction or in terms of the abundance of the reacting ion, P^+. In the second, the relative abundances of two metastable peaks due to competitive reactions from the same reacting ion are employed. While metastable ion abundances are usually relatively insensitive to the internal energy distribution resulting from electron impact, fragment ion and parent ion abundances are strongly dependent upon $P(\epsilon)$. Hence, the quotients $[m*]/[P^+]$ and $[m*]/[F^+]$ have proved of only limited value in characterizing gaseous ions. Indeed, for a given ion, these quotients provide a measure of the average internal energy of the reacting ion and an application to the characterization of internal energies is to be found in the study of the degrees of freedom effect upon fragment ion reactivity (see Appendix I).

Some dependence upon internal energy is also found[221] when the second method, the abundance ratio of two competitive transitions from the metastable ion, is considered. This effect, however, is small. If reactant ions of unknown structure are examined and a similar abundance ratio is found, this is evidence that the ions have the same structure. If the ions are structurally different, a large difference in the abundance ratio can be expected, often to the extent that only one reaction is observed. Some examples of the use of this quotient appear in Chapter 6.

Isotope effects

One area in which the consequences of the generally small excess energies of metastable ions are particularly interesting concerns isotope effects upon metastable peak heights. Assuming that there are no significant differences in the collection efficiencies of $((M-H)^+$ and $(M-D)^+$ ions, the ion abundances provide a direct measure of the relative extents of H· and D· loss from a partially deuteriated molecular ion. Thus, the isotope effect k_H/k_D is given by the relative abundances of the $(M-H)^+$ and $(M-D)^+$ product ions. It is important to realize that since we are dealing with non-equilibrium isotope effects in isolated ions, k_H/k_D values may be very large indeed. Very high values of this ratio have been reported, including the extreme case where D· loss could not be detected while H· loss could.

Figure 57 illustrates a potential energy diagram for H· and D· loss from equivalent positions in the same partially deuteriated molecular ion. The curve is symmetrical about the stable configuration of the ion and the only differences arise from the zero-point energy terms. Since H· and D· are being lost from a single molecular ion, there is only a single zero-point energy for this species. The activated complexes leading to H· and D· loss possess zero-point energy terms associated with all the degrees of freedom of the ion except that involved in the reaction under consideration. The activated complex corresponding to H· loss must have the lower zero-point energy since

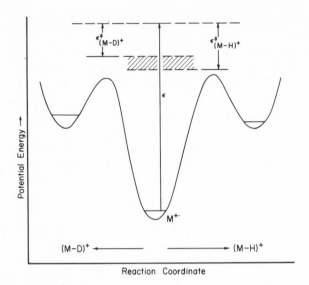

Fig. 57. Potential energy diagram for loss of H$^{\bullet}$ for D$^{\bullet}$ from a molecular ion M$^{+\bullet}$ of internal energy ϵ. Because of the differences in zero-point energies associated with C—H and C—D bonds, the internal energy, $\epsilon^{\ddagger}$, of the activated complex for H$^{\bullet}$ loss is greater than that for D$^{\bullet}$ loss.

it possesses a C—D bond not involved in the fragmentation step. This confers a lower zero-point energy on the ion than is provided by the C—H bond retained in the complex undergoing D$^{\bullet}$ loss. Hence, an ion with internal energy ϵ (Fig. 57) has a larger excess energy when it adopts the configuration appropriate to H$^{\bullet}$ loss than if it adopts that for D$^{\bullet}$ loss. The activation energy for H$^{\bullet}$ loss is, therefore, lower than that for D$^{\bullet}$ and consequently, this reaction is faster, leading to $k_H/k_D > 1$. The hatched area of the figure covers the range of internal energies for which an ion might undergo H$^{\bullet}$ loss, but could not undergo D$^{\bullet}$ loss at all. If the only ions sampled were those having energies in this range, then the isotope effect would be infinite. It can readily be seen from the figure that as the internal energy ϵ increases, the ratio of $\epsilon^{\ddagger}_{(M-H)^+}$ to $\epsilon^{\ddagger}_{(M-D)^+}$ will decrease, and k_H/k_D will approach unity. In agreement with this, numerous instances have been encountered in which isotope effects for reactions occurring in the ion source have been found to be smaller, and sometimes very much smaller, than those found for metastable ions. This type of experimental data concerning isotope effects on metastable ion abundances supports the conclusion that the average excess energy of metastable ions is low, but in the absence of reliable k

versus ϵ data, this approach cannot yet be used to obtain quantitative information on $\epsilon^{\ddagger}$ values.

If two different isotopically substituted molecular ions are examined for H· and D· loss, the situation becomes more complex because the internal energy distributions of the two molecular ions will not be identical and each will have its own zero-point energies. Nevertheless, in this case also, experimental results show that $\epsilon_{0(M-H)^+} < \epsilon_{0(M-D)^+}$ and $k_H/k_D > 1$.

It is striking that k_H/k_D for H· loss from metastable ions, determined using the same instrument and the same experimental conditions, varies from 1.9 for benzene to > 1000 for isobutane[31]. This difference appears to be related directly to the magnitude of $\epsilon^{\ddagger}$ for the activated complex in question. Hence, if the reaction for H· (and D·) loss from a given molecular ion has a slow rise of k with ϵ, then the ions fragmenting in the time interval which corresponds to the observation of metastable peaks will have relatively high internal energies compared with those for reactions in which k increases rapidly with ϵ. This effect is analogous to that responsible for the smaller isotope effects observed for reactions in the ion source compared with those for metastable ion reactions. It is illustrated in Fig. 58 which shows the k *versus* ϵ curves for H· and D· loss in reactions with slow and fast rises of k with ϵ. The magnitude of the isotope effect may be independent of the difference in the activation energies involved. Furthermore, since the slopes of these curves are

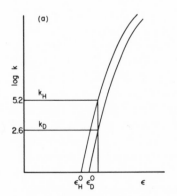
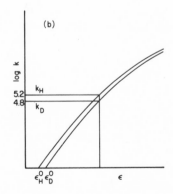

Fig. 58. Effect of entropy of activation upon the isotope effect k_H/k_D. In (a), the rate constant increases rapidly with internal energy giving a large isotope effect, while in (b), the rate constant increases slowly giving a relatively small isotope effect.

related to the entropy of activation for the process, the generalization can be made that reactions in which $H^{\cdot}$ is lost by a low-frequency factor process (including rearrangements) will be associated with small k_H/k_D values and *vice versa*.

One further conclusion can be drawn from this data. Since k_H/k_D can vary a great deal with the average internal energy of the reactant ion, especially if a process with a rapid rise of k with ϵ is chosen, experimental k_H/k_D values could be used to characterize the time scale appropriate to fragmentation of metastable ions. Comparison between different instruments should become possible.

RELEASE OF KINETIC ENERGY

Effects of reverse activation energy and of excess internal energy

The conversion of internal energy to translational energy of the products in the course of the fragmentation of a metastable ion is covered in this section. Coverage is limited, however, to processes which involve the accumulation of vibrational energy in appropriate modes to surmount an energy barrier. Tunneling and predissociation are discussed in a later section and the kinetic energy release accompanying these phenomena is considered there.

The kinetic energy, T, released in the decomposition of a metastable ion can arise[163, 255] from two separate sources: (*i*) the excess energy, $\epsilon^{\ddagger}$, of the activated complex which is available for partitioning between the internal energies of the products and translational energy of their separation and (*ii*) the reverse activation energy, $(\epsilon_0{}^r)$, which is also partitioned between internal and translational energy. Hence, we may write

$$T = T^e + T^{\ddagger} \tag{37}$$

where T^e is the contribution to the kinetic energy released from the reverse activation energy and $T^{\ddagger}$ is that from the excess energy of the activated complex. T^e depends upon the detailed energetics and dynamics of the reaction and is intrinsic to the particular process; $T^{\ddagger}$ varies with the internal energy of the reactant ion and has no fundamental significance. Figure 59 summarizes these relationships. Although the separation of T^e and $T^{\ddagger}$ cannot be justified rigorously, it is possible to determine T for various values of $\epsilon^{\ddagger}$ (including small values) and the distinction, therefore, is phenomenologically valid.

Early results on kinetic energy releases, especially the correlation

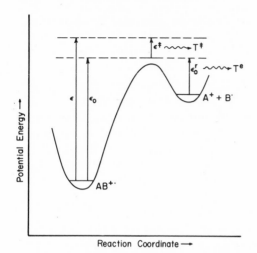

Fig. 59. Contribution of the non-fixed energy of the activated complex and of the reverse activation energy to the observed kinetic energy release. Some portion of $\epsilon^{\ddagger}$ will appear as kinetic energy $T^{\ddagger}$, and some portion of ϵ_0^{r}, will appear as kinetic energy T^{e}.

observed between the energy release for NO˙ loss in p-substituted nitrobenzenes and the electron-donating power of the substituent[75], led to the view that the entire reverse activation energy is released as kinetic energy. An early indication that this was not the case was provided by the loss of C_2H_4 from phenetole, a reaction for which a reverse activation energy of more than one electron volt was estimated but which, nevertheless, was associated with a narrow metastable peak[277]. Although the structure of the products and, hence, the magnitude of the reverse activation energy might be questioned in an isolated example such as this, extensive studies (see below) have confirmed the principle that some fraction of ϵ_0^{r}, dependent on the precise electronic course of the reaction, will appear as kinetic energy. It may also be worth emphasizing that the reverse activation energy is certainly not equipartitioned between the dissociation mode and the other degrees of freedom. That this is so may be seen by considering the case of a release of 0.2 eV for a positive ion having 50 oscillators. This would, according to equipartition, necessitate a reverse activation energy of 10 eV, which is much larger than activation energies. And even if only one fifth of the oscillators were effective, the energy required would still be too large.

Because of the composite origin of T expressed in eqn. (37), interest focuses on two extreme situations, $T \cong T^{\ddagger}$ and $T \cong T^e$. If, in any practical system, either of these approximations can be justified, then experimentally determined kinetic energy releases can be related to either the excess internal energy of the ion, or to the reverse activation energy. Both T^e and $T^{\ddagger}$ are quantities of very considerable interest, since they can, through knowledge of the individual energy partitioning relationships, lead back to $\epsilon_0^{\,r}$ and $\epsilon^{\ddagger}$. Both these quantities are important in thermochemical determinations by mass spectrometry and $\epsilon_0^{\,r}$ is, in addition, a characteristic of the reaction in question, and hence of potential value as an indicator of ion structure.

Reactions which can be expected to have substantial reverse activation energies include rearrangements in which a stable neutral molecule and a stable ionic product are generated. By way of contrast, simple bond cleavages are expected to have zero or very small reverse activation energies: ion/free-radical recombinations generally proceed with zero activation energies. The internal energies of metastable ions are small (see above) and for large molecules the fraction $T^{\ddagger}/\epsilon^{\ddagger}$ is expected to be small[163,263]. Hence $T^{\ddagger}$ will be small, but not negligible. This conclusion is based on experimental evidence[220,255] which has been interpreted as showing that $\epsilon^{\ddagger}$ is statistically partitioned over the available degrees of freedom of the activated complex including that ultimately associated with product separation. Agreement between the kinetic energy released in reactions in the ion source and the calculated release based on statistical partition is often good, except that some rearrangement and elimination reactions give much larger releases than expected. A reverse activation energy is assumed to be responsible for these deviations[220,263].

Figure 59 illustrates the dual origin of T in the case of a process $(AB^{+\cdot} \rightarrow A^+ + B^\cdot)$ in which the reverse activation energy is appreciable. In order to compare the relative contributions of $T^{\ddagger}$ and T^e in different reactions, it is instructive to compare the energy released by metastable ions with that released by the higher-energy ions which fragment in the ion source[156]. When this is done, an interesting fact emerges: fragmentation reactions which involve bond formation and which can, therefore, be expected to have sizeable reverse activation energies, release approximately the same energy whether the process is studied in the ion source or in the field-free region. Propane provides a typical example: the elimination of H_2 from this molecular ion should have a reasonably large reverse activation energy and it is found that the energy released, measured by the less accurate deflection technique for ion source reactions, is 0.65 eV while the value found for metastable ions is 0.40 eV, proportionately not much less. The simple loss of H$^\cdot$ from the same molecular ion gives an energy release of 0.18 eV in the ion source, but only 0.002 eV in the field-free region. While it could be argued that reactant ions in a different electronic state were sampled in the field-free region in the second reaction, the result has general validity and supports the view that reactions in which $\epsilon_0{}^r$ is small or zero will give energy releases which depend almost entirely upon $\epsilon^\ddagger$.

These observations suggest that thermochemical data can be corrected for the excess energy term $\epsilon^\ddagger$ provided $\epsilon_0{}^r$ is small. Haney and Franklin[126] adopted this procedure and assumed that $\epsilon_0{}^r$ was *always* negligible. As later results will show, this is not always the case. It does seem reasonable, however, to compare the kinetic energy release in the ion source with that in the field-free region and so to determine those cases in which $\epsilon_0{}^r$ can be ignored. For these, the appearance potential (measured for reactions occurring in the source) can be corrected for the energy release accompanying the reactions. The considerable improvement in ionic enthalpies which results from such a correction can be illustrated using one of Haney and Franklin's examples. These authors used an empirically determined relationship between T and ϵ (in our terminology $T^\ddagger$ and $\epsilon^\ddagger$)

$$T^\ddagger = \frac{\epsilon^\ddagger}{\alpha s} \tag{38}$$

where s is the number of oscillators in the reactant ion and α is an arbitrary parameter taken as 0.44. It is apparent that this relationship is based on statistical partitioning of $\epsilon^\ddagger$ in agreement with expectation. For example, the heat of formation of HCO$^+$ generated from formic acid is found to be 203 kcal.mole^{-1}, and the ion generated from acetaldehyde is found to have a heat of formation of 220 kcal.mole^{-1}, neither value being corrected for excess energy. The kinetic energy released in the ion source reaction

$$\text{HCOOH}^{+\cdot} \rightarrow \text{HCO}^+ + \text{OH}^\cdot$$

as determined from the peak shape using time-of-flight analysis, was

1.1 ± 0.5 kcal.mole^{-1} (0.048 eV) from which $\epsilon^{\ddagger}$ is estimated from eqn. (38) as 4 ± 2 kcal.mole^{-1} leading to a corrected enthalpy of formation of 199 ± 2 kcal.mole^{-1}. The energy release accompanying the second reaction

$$CH_3CHO^{+\cdot} \rightarrow HCO^+ + CH_3{}^{\cdot}$$

was 3.5 ± 0.6 kcal.mole^{-1} (0.152 eV) from which $\epsilon^{\ddagger}$ is estimated as 23 ± 4 kcal.mole^{-1}. This leads to a corrected enthalpy value of 197 ± 4 kcal.mole^{-1}. The agreement between these two values is typical of the improvement achieved by proper correction for excess energy terms.

As an alternative method of correcting for $\epsilon^{\ddagger}$, the enthalpy can be determined from the appearance potential of the product from the metastable ion. There will be a small contribution from $\epsilon^{\ddagger}$ in this case, but it can be estimated from the kinetic energy release using the above procedure. For example, in an ion with 10 atoms, there are 24 degrees of freedom of which perhaps 10 will be active. Hence, if $T = 10$ meV (a common range for reactions which do not involve rearrangement), $T^{\ddagger} \leqslant 10$ meV and $\epsilon^{\ddagger}$ is likely to be less than 100 meV. Hence, the enthalpy determined from the appearance potential of the product of the fragmentation of the metastable ion in such a case is not likely to be in error (low) by more than 2 kcal.mole^{-1}.

The correlation which can be observed between large kinetic energy release and the occurrence of a rearrangement or elimination reaction, further emphasizes that for metastable ions, $\epsilon^{\ddagger}$ is small (exceptional cases involving extremely low entropies of activation are excluded). This conclusion also emerges from the comparison of kinetic energy releases in metastable ion and ion source reactions. This is an important result which will be used later and is worth further discussion. The quantity of interest is related to the kinetic shift for metastable ions. As already shown, this will be smaller than the kinetic shift for ion source reactions. This allows us to put an approximate upper limit on $\epsilon^{\ddagger}$ for any particular reaction of a metastable ion. Calculated k versus ϵ data for some typical reactions of organic ions show values of a few tenths of an electron volt for this quantity[197]; other calculations give even smaller values[263].

The above considerations also show that when the kinetic energy release measured for reactions of metastable ions is large (> 0.1 eV) these data may, with reasonable assurance, be related to $\epsilon_0{}^r$ only. That is to say, the approximation $T \cong T^e$ is then valid. Given this approximation, it merely requires some knowledge of the $T^e/\epsilon_0{}^r$

relationship to make significant corrections to thermochemical data based on appearance potentials. This type of information has been obtained only to a limited extent, but already it seems clear that the energy partitioning quotient T^e/ϵ_0^r varies with the entropy of activation. Some results are given below and others appear in Chapter 6. It is significant that considerable improvement can be achieved in the accuracy of determining ionic heats of formation (see below) even using the very approximate corrections at present possible.

As just noted, the approximation $T \cong T^e$ is valid if T is large. Now ϵ_0^r must be large for T to be large, although the converse does not hold. Hence, an experimental route to ϵ_0^r is required in order to study the partitioning of the reverse activation energy. One approach is illustrated in Fig. 60. The quantity ϵ_{excess}, the sum of ϵ_0^r and $\epsilon^{\ddagger}$,

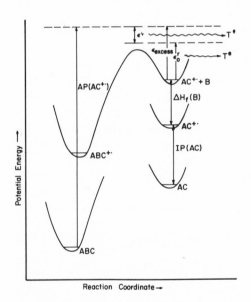

Fig. 60. Experimental determination of ϵ_{excess} $(\epsilon_0^r + \epsilon^{\ddagger})$ for a reaction $ABC^{+\cdot} \rightarrow AC^{+\cdot} + B$.

is experimentally accessible, and for metastable ions, $\epsilon^{\ddagger}$ is small. Thus, in the cases we are considering where ϵ_0^r is large, the approximation $\epsilon_{excess} \cong \epsilon_0^r$ is valid. It is also worth noting that the approximation

$$\frac{T}{\epsilon_{excess}} \cong \frac{T^e}{\epsilon_0^r}$$

should be better than the individual approximations $T \cong T^e$ and $\epsilon_{excess} \cong \epsilon_0{}^r$ since both the experimentally accessible quantities represent upper limits to the quantities of interest.

As shown in the figure, ϵ_{excess} is measured as the difference between the energy of the activated complex, including the internal energy $\epsilon^{\ddagger}$, and the enthalpy of the reaction products in their ground states. The heat of formation of the fragment ion is determined by direct ionization of the neutral species (AC), that of the neutral product (B) is known from thermochemical measurements or estimated by the group equivalent method. Thus, formally, for the reaction

$$ABC^{+\cdot} \rightarrow AC^{+\cdot} + B$$

$$\epsilon_{excess} = AP(AC^{+\cdot}) - IP(AC) + \Delta H_f(ABC) - \Delta H_f(AC) - \Delta H_f(B) \quad (39)$$

where IP(AC) is an adiabatic ionization potential leading to an unexcited $AC^{+\cdot}$ ion and the only excess energy term involved in $AP(AC^{+\cdot})$ is the kinetic shift.

In applying this equation certain assumptions can be made.

(i) Measurements need not refer to $0°K$. Rather, enthalpies may refer to $298°K$ and ionization and appearance potentials to a higher temperature. This is justified by the fact that the enthalpy difference $\Delta H_f(ABC) - \Delta H_f(AC) - \Delta H_f(B)$ appears in eqn. (39) as does the difference $AP(AC^{+\cdot}) - IP(AC)$.

(ii) Since $AP(AC^{+\cdot})$ is determined for ion source reactions, it is likely to be in error by the kinetic shift. Any differences between $\epsilon^{\ddagger}$ for metastable ions and for ions fragmenting in the source at threshold will, however, have a negligible effect on ϵ_{excess} in the reactions under consideration where $\epsilon_0{}^r$ makes the overwhelming contribution to ϵ_{excess}.

(iii) It is assumed that ionization of AC gives unexcited ground state $AC^{+\cdot}$ ions. This will usually be the case to a good approximation for aromatic systems and any small error will tend to decrease ϵ_{excess} and hence to increase the accuracy of the approximation $\epsilon_{excess} \cong \epsilon_0{}^r$.

(iv) The above assumptions are in the nature of simplifications to facilitate the experimental determination. In contrast, the assumption that the ion formed by the fragmentation of $ABC^{+\cdot}$ is structurally identical to that generated by ionization of AC is essential to the entire treatment and if wrong would lead to entirely erroneous conclusions. This assumption must be considered separately for each individual reaction. In some cases, the nature of the product is known with confidence from other methods, but in others this is not the case and the energy-partitioning method can be used to deduce reaction mechanisms using available data on the energy-partitioning quotient. This type of application is still in its infancy, but appears to have considerable potential (see Chapter 6).

It may be worthwhile to illustrate the study of energy partitioning by considering a case in which there is independent evidence as to the nature of the product ion. The loss of formaldehyde from the molecular ion of anisole and substituted anisoles provides a good example. Besides illustrating energy partitioning, this example also introduces a new feature of metastable peak shapes which can only be detected under conditions of high energy resolution: the occurrence of superimposed peaks of different widths. We shall refer to these as composite metastable peaks and an example is given in Fig. 61. It is important to note that both peaks are due to unimolecular transitions involving parent and daughter ions of the same nominal mass. In many cases, these ions have the same empirical formulae and only differences in ion structure are responsible for the different kinetic energy releases which cause the differences in metastable peak widths.

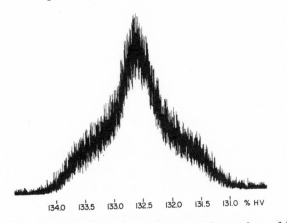

134.0   133.5   133.0   132.5   132.0   131.5   131.0   % HV

Fig. 61. Composite metastable peak due to loss of H_2CO from p-methylanisole molecular ions by two competitive unimolecular mechanisms. The HV scan method was used to record this peak.

There is evidence from previous studies that the loss of formaldehyde from anisole and from substituted anisoles occurs through a four-membered cyclic transition state, as shown[18,192].

$$Y \longleftrightarrow \text{(ring)} \underset{H}{\overset{O}{\diagdown}} CH_2 \quad \Big]^{+\cdot} \longrightarrow \quad Y \longleftrightarrow \text{(ring)} \quad \Big]^{+\cdot} + H_2CO$$

There is also evidence from isotope labeling studies[165] which supports the five-membered cyclic hydrogen transfer.

These mechanisms are not mutually exclusive, however, and it will be seen that the energy partitioning results suggest that both occur.

The metastable peak for loss of mass 30 from the molecular ion is composite for all substituted anisoles and except in the case of the nitroanisoles, where loss of $NO^{\bullet}$ and loss of H_2CO give a product ion of the same mass, this indicates that two separate processes contribute to the formaldehyde elimination reaction. The energy release, strictly the average energy release, for each of the contributing reactions was measured by deconvolution of the composite metastable peak. Table 1 summarizes this data for a series of p-substituted anisoles. It includes two values of the kinetic energy release, corresponding to the narrow and the broad components of the metastable peak and denoted T_s and T_1, respectively. The relative abundances of these two components are also entered (in parentheses). Other columns give the ionization potential of the anisole, IP(M), the appearance potential of the $(M-H_2CO)^{+\bullet}$ fragment ion, AP(F^+), and the ionization potential of the corresponding monosubstituted benzene IP(F). Standard heats of formation of the anisole and the substituted benzene at 298°K are also given in each case.

These thermochemical data allow the calculation of ϵ_{excess} for the reaction in which the product of the anisole fragmentation is the corresponding ionized benzene derivative. In the alternative reaction, the product is the less stable carbene. Now, the larger energy release, T_1, can be associated with formation of the more stable product, the ionized benzene, hence T_1/ϵ_{excess} is the energy partitioning parameter for formaldehyde elimination by the four-centered cyclic rearrangement. It is striking that this quotient (Table 1) is independent of substituent within experimental error. Such a finding would be unlikely if any of the assumptions in the energy-partitioning treatment were grossly in error and this supports the assignment of T_1 to

TABLE 1

Energetics for loss of formaldehyde from the molecular ions of p-substituted anisoles[a]

Substituent	$\bar{T}$	T_s	IP(M)	AP(F$^+$)	IP(F)	ΔH_f°(M)	ΔH_f°(F)	ϵ excess	$\bar{T}_l/\epsilon_{excess}$
H	0.32 (60%)	0.02 (40%)	8.2_0	11.5_0	9.2_1	-0.69	$+0.86$	1.9_6	0.17
CH$_3$	0.36 (70%)	0.02 (30%)	7.8_5	11.2_3	8.6_7	-1.00	$+0.52$	2.2_6	0.16
Cl	0.34 (>95%)	b (>5%)	8.1_8	11.4_2	8.9_9	-1.05	$+0.55$	2.0_6	0.17
Br	0.28 (85%)	0.02 (15%)	8.3_9	11.5_2	8.9_8	-0.53	$+1.04$	2.1_2	0.14
CN	0.36 (40%)	0.04 (60%)	8.7_4	12.3_9	9.7_7	$+0.66$	$+2.31$	2.1_8	0.17

[a] All energy terms are given in eV.
[b] Abundance too low to allow determination.

the four-centered cyclic rearrangement mechanism. Further support comes from substituent effects on relative metastable peak areas (see below).

In order to interpret the mechanism of composite metastable peak formation, several experiments were done using p-methylanisole, chosen because the two metastable peaks are of comparable abundance. First, the pressure in the first field-free region of the mass spectrometer was varied. The signal was unchanged in shape, neither contributing process being pressure-enhanced. Second, the effects upon the metastable peak shape of lowering the ionizing electron energy and of raising the source temperature were investigated. No changes were observed. Finally, the anisole molecular ion was generated by fragmentation of the corresponding O-methyl oxime ether: it behaved identically to the ion generated by direct ionization. Analogous results were also obtained for the anisole molecular ion generated by direct ionization and by fragmentation of dimethoxybenzene.

Since neither of the two formaldehyde elimination reactions contributing to the composite peak is collision-induced, it remains to be established whether two forms of the reactant, of the ionic product or of the neutral product are responsible for the composite peak.

The first excited state of formaldehyde, the neutral product in the reaction under consideration, has an energy of 76 kcal.mole^{-1} (3.3 eV), which is too high to allow the possibility that formation of the ground and first electronically excited states of the neutral are responsible for the composite peak. Formation of either the neutral or the ionic product in different vibrational states can be dismissed because it implies that composite metastable peaks should be the rule, not the exception, and because the observation of only two vibrational states is unreasonable. These considerations imply that the composite nature of the $M^{+ \cdot} \rightarrow (M-CH_2O)^{+ \cdot}$ transition must be due either to the generation of the ionic product in two different electronic states from a single reactant ion or to the reaction of two isolated electronic states of the reactant ion. In distinguishing between these possibilities, it must be emphasized that the former does not exclude the possibility that one or both of the reaction channels will involve intermediates, but it does require that they be in equilibrium with the reactant.

The fact that the relative contributions of the two peaks do not change detectably with electron energy indicates that if two different electronic states of the reactant ion are involved, their energies are similar. In addition, the ionization efficiency curves for the

$(M—H_2CO)^{+\cdot}$ ion showed no breaks or other evidence for two processes with measurably different activation energies. The results of the experiments using different sources of the reactant ion (ionization *versus* fragmentation) make it unlikely that H_2CO loss occurs from two reactants in isolated electronic states. The results on a competitive reaction, methyl loss from the molecular ion, also make unattractive the suggestion that there exist isomeric forms of the reactant which are not in equilibrium with each other. If isolated electronic states were involved, composite metastable peaks would be likely for $CH_3^{\cdot}$ loss in at least some of the p-substituted anisoles. p-Methylanisole provided the only instance in which a composite peak for methyl loss was observed; this is a special case in which the composite peak has an entirely different origin due to the possibility of ring expansion[202] accompanying methyl loss.

Hence, metastable molecular ions of anisoles apparently undergo H_2CO elimination by two reaction channels from the same reactant with formation of different isomers of the product ion. If it is assumed that these are the four-membered and five-membered hydrogen transfer reactions for which independent evidence is available, then the validity of this postulate can be tested using the thermochemical and substituent effect data accrued in this study. It should be noted that the measured appearance potential for the $(M—H_2CO)^{+\cdot}$ ion can be used in deriving energy-partitioning data in spite of the fact that two reactions are involved. This is justified in the light of the evidence already presented which shows that the two reactions have similar activation energies.

The difference in the kinetic energies released in forming the two components of the composite metastable peak is approximately 300 meV. This implies that the difference in the enthalpies of the product ions must be in excess of 300 meV. This seems entirely reasonable in view of the formulation of the less stable product as the ionized carbene.

The thermochemical data are, therefore, consistent with the suggestion that the proposed mechanisms of hydrogen rearrangement are indeed involved in formaldehyde loss from substituted anisoles. The experimental results also provide another means of testing this hypothesis. The relative proportions of the processes leading to large and small energy release follow this generalization: electron-donating substituents increase the relative contribution of the process associated with the larger value of T, electron-withdrawing substituents decrease it. Substituents also show very little effect upon the magnitude of either the larger or the smaller energy release.

These facts are all accommodated if the process leading to the larger value of T involves the four-centered and that leading to the smaller value involves the five-centered hydrogen transfer mechanism and if both hydrogen transfers occur to radical sites. A detailed analysis of these substituent effects appears in the original article[89].

In addition to the importance of energy partitioning data as a new source of information on reaction mechanisms and reaction dynamics, it has considerable thermochemical significance. If the enthalpy of, say, the chlorobenzene ion $C_6H_5Cl^{+\cdot}$ were determined from the appearance potential of this ion as generated from p-chloroanisole, the excess energy term arising from the reverse activation energy would be a far more serious source of error than the kinetic shift term. The total excess energy would be approximately six times the measured kinetic energy release accompanying the fragmentation. An excess energy term of more than 2 eV would have to be taken into consideration. The thermochemistry and reaction dynamics for formaldehyde loss from anisole are summarized in Fig. 62.

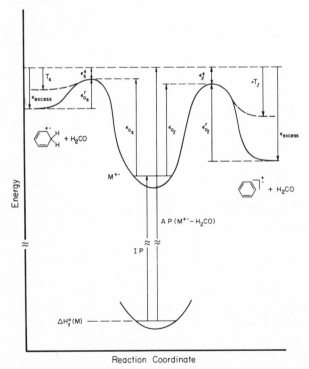

Fig. 62. Thermochemistry pertaining to H_2CO loss from anisole by four- and five-centered hydrogen rearrangements.

The general approach to $T^e/\epsilon_0{}^r$ just outlined has been applied to several series of substituted compounds and it seems that the energy partitioning quotient is related to the entropy of activation — the higher the entropy of activation, the greater the fraction of the reverse activation energy released as kinetic energy. This is a reasonable result, since it means that in tight activated complexes, the energy made available as the system moves towards the product configuration is not easily absorbed in internal modes, these being either frozen out or of decreased amplitude. Some rearrangement reactions, including NO˙ loss from nitroaromatics, occur with release of virtually all the available energy. In this case, the major process involves a three-centered cyclic rearrangement. Elimination of ethylene from phenetoles and other reactions involving transition states of larger ring size give far smaller energy-partitioning parameters. The importance of the energy partitioning results for the understanding of ion structures is treated in more detail in Chapter 6.

The effective transfer of energy from the loci at which it first appears as the system moves from the activated complex configuration towards products is expected to involve two basic requirements. These are (i) effective interaction of the degree of freedom represented by the reaction co-ordinate with other oscillators and (ii) sufficient time for this interaction to occur. Since the total number of oscillators that are effective is reduced in a tight activated complex, energy transfer becomes less effective. This means that more energy is retained at the loci at which it appears, $i.e.$ in the form of translational energy, since movement along the reaction co-ordinate represents fragmentation. It is perhaps worth noting that the experimental correlation with entropy of activation is more fundamentally a correlation with the entropy of activation for the reverse reaction. The two correspond in reactions where ϵ_0 and $\epsilon_0{}^r$ are both appreciable, but it is oscillator interaction in the region of the potential energy surface over which the activated complex moves towards products which is significant. The second of the two basic factors controlling $T^e/\epsilon_0{}^r$ is the time spent by the system in moving from activated complex to products. This time is, in part, dependent upon the kinetic energy release itself, hence the net effect is complex. The principle involved may, however, be illustrated in simple terms.

Consider the case of an ion $R_1R_2{}^{+\cdot}$ of mass-to-charge ratio 100 breaking into fragments $R_1{}^+$ and $R_2{}^\cdot$ each of mass 50 units. If the initial kinetic energy of $R_1R_2{}^{+\cdot}$ corresponds to an accelerating voltage of 10 kV and the measured energy release corresponds to 2 eV, the fragments separate with a relative velocity of 4×10^5 cm/sec as discussed elsewhere. If we assume that the effective distance over which the separating particles can interact is of the order of 10Å, then the interaction time is only of the order 2×10^{-13} sec. Thus, as soon as separation commences in the reaction co-ordinate, the dissociation proceeds rather more as a "vertical" than an "adiabatic" process and the possibility of converting internal into external energy is limited. Indeed, the total interaction time in this rather extreme case is so small that it is comparable to a vibrational period and this will severely circumscribe the extent to which energy can be distributed and will, therefore, maximize the value of $T^e/\epsilon_0{}^r$.

In addition to the control upon $T^e/\epsilon_0{}^r$ exerted by oscillator coupling and reaction time, as just discussed, it must be remembered that the release of energy in translational form is always favored from the point of view that this maximizes the entropy of the system, since the number of translational states is effectively infinite.

Energy release as an ion structure characteristic

The foregoing discussion has covered the dependence of the kinetic energy release, T, upon the reverse activation energy of the reaction and upon the excess energy of the reactant ion. The usefulness of T as a means of characterizing ions can now be discussed. If T is determined from metastable peak widths, then the fact that the mass spectrometer detects as metastable ions only those which fragment in a specified narrow time interval means that the excess internal energies, $\epsilon^{\ddagger}$, are quite closely fixed by the instrumental conditions. Hence, whether or not $\epsilon_0{}^r$ makes a major contribution to T, the kinetic energy released will be far less sensitive to internal energy and more sensitive to ion structure than are other parameters of metastable ions, including their abundances. Thus, by comparison of ions formed from different sources, for example by fragmentation of different molecular ions, it should be possible to determine from the kinetic energies released in their further fragmentation whether or not they are identical, even if they do not have the same internal energy distributions. To illustrate, the energy release for the reaction $C_{12}H_{10}O^{+\cdot} \rightarrow C_{11}H_{10}{}^{+\cdot} + CO$ is 0.435 eV when the molecular ion of diphenyl ether is examined and 0.438 eV when the $(M-CO_2)^{+\cdot}$ ion of diphenyl carbonate is examined. In another case, $C_6H_5{}^+ \rightarrow C_4H_3{}^+ + C_2H_2$, the release is 0.018, 0.024 and 0.025 eV when the reactant ion is generated from benzene, anisole and styrene, respectively. It is found that when ions of identical structure are generated by two routes, one of which involves direct ionization and the other an intermediate fragmentation step, then the slightly larger energy release is usually associated with the ion formed by way of fragmentation. This may be the result of the shorter interval between fragment ion formation and arrival in the field-free region of the mass spectrometer where the relevant ionic reaction is sampled. Very much larger differences in the values of T arise, however, if the ion structures are not identical. To illustrate the magnitude of the differences which are possible in these cases, acetophenone and butyrophenone release 0.008 and 0.053 eV, respectively, in the reaction $C_8 H_8 O^{+\cdot} \rightarrow C_7 H_5 O^+ + CH_3^{\cdot}$. This disparity is thought to be associated with structural differences in the reacting ions[155].

Another phase of the enquiry into effects of ion preparation upon kinetic energy release is concerned with the possible existence of a "degrees-of-freedom" effect upon T, analogous to that observed upon ion abundance (see Appendix I). In the ion abundance study, a given ion is generated from several members of an homologous series and the metastable peak associated with its further fragmentation is found to decrease in abundance as the size (number of degrees of freedom) of the neutral homolog is increased[190]. The effect has been shown[171] to be due to a decrease in the average internal energy of the specified fragment ion as the series is ascended. The metastable ion $C_2 H_5 O^+$ which undergoes competitive fragmentations to give $H_3 O^+$ and CHO^+ was chosen for investigation since it was also the subject of an ion abundance study. No variation in kinetic energy release with ion origin could be detected; 2-propanol and 2-octanol give $C_2 H_5 O^+$ ions which release 24 and 25 meV, respectively, for the reaction $C_2 H_5 O^+ \rightarrow H_3 O^+ + C_2 H_2$ and values of 860 and 830 meV for the competitive reaction, $C_2 H_5 O^+ \rightarrow CHO^+ + CH_4$. The corresponding abundance ratios, $[m^*]/[P^+]$, differed by a factor of about 20 in each of these reactions, again emphasizing the relative insentivity of T to variations in the internal energy distribution of the fragmenting ion.

The foregoing results establish the general usefulness of T as a probe into ion structure and indicate the type of variation which can be expected to result from differences in methods of preparation, vis-à-vis those which result from differences in ion structure.

The reactions of metastable ions occur from a narrow band of internal energies selected from the broad distribution of energies which results from electron impact. Changes in source temperature and in the electron beam energy, therefore have little effect upon the measured kinetic energy release. No instance of a variation in T with electron energy has yet been reported although it may be noted that some of the instruments in which T can most accurately be measured are equipped with "bright ion" sources which are unsuitable for studies at low electron energy. In one study[155] the effect of source temperature upon T was followed for two reactions, $M^{+\cdot} \rightarrow (M-C_2H_2)^{+\cdot}$ in benzene and $M^{+\cdot} \rightarrow (M-CH_3)^+$ in acetophenone over the range $120-265°C$. In the benzene reaction, T increased steadily, but only from 26 to 29 meV, while in the acetophenone case, it increased from 8.3 to 9.6 meV. These results demonstrate the insensitivity of the measured value of T to quite large variations in the internal energy distribution of the reactant ions examined. Changes in the time of observation of metastable ions, such as can be produced by changes in ion-accelerating voltage, by varying the source residence time and by observing reactions in another field-free region, are more effective in changing the kinetic energy release[40]. Even these methods, however, are limited by the fact that the internal energies of metastable ions are very small and that the kinetic energy released in many reactions is primarily dependent upon the reverse activation energy not upon the internal energy of the reactant ion.

Isotope effects

Isotope effects operate upon kinetic energy release and are a promising technique for extracting more information on reaction energetics. The limited data available shows that the effects may be large. For example, the reaction $H-\overset{\cdot}{C} = \overset{\cdot}{O}-H \rightarrow CHO^+ + H^{\cdot}$ in methanol is accompanied by an energy release of 0.19 eV, while the corresponding loss of $D^{\cdot}$ from the O-deuteriated ion ($H-\overset{\cdot}{C} = \overset{+}{O}-D$) has an energy release of 0.39 eV (ref. 39). Similarly, the energy release for HCN loss from the metastable pyrimidine molecular ion is 0.9 meV while that for DCN loss in the corresponding reaction of d_4-pyrimidine is 2.1 meV (ref. 54). No general interpretation of these and other isolated results is available. However, $H^{\cdot}$ and $D^{\cdot}$ loss from a series of partially deuteriated ions has been studied and in most cases the ratio of the kinetic energy releases was close to unity[30]. The results can be understood by referring back to Fig. 58, which illustrates the energetics applicable to reaction of a molecular ion which can lose either $H^{\cdot}$ or $D^{\cdot}$ by simple $C-H(D)$ bond cleavage. The potential energy surface is symmetrical about the molecular ion configuration and only differences in zero-point energies are of concern. The simplification is made that the internal energy distribution appropriate to fragmentations in the field-free region can be represented as a single discrete value. It is recognized that the excess energies for the two reactions might actually differ marginally if sampling over the appropriate distribution were different for the two cases. The two sets of products $(M-H)^+ + H^\cdot$ and $(M-D)^+ + D^\cdot$ differ only in zero-point energies of the ions, the former being lower by approximately 1 kcal.mole^{-1} (estimated from the bond vibrational frequencies). The two activated complexes differ in that one (that in which $H^\cdot$ is lost) retains a C—D bond, the other retains a C—H bond. These particular bonds, whether C—D or C—H, may be a little stronger or a little weaker in the activated complex than in the products and, hence, will tend either to make the difference in zero-point energies of the two activated complexes slightly greater than or slightly less than the difference in zero-point energies of the two sets of products. Hence ϵ_0^r $(M-H)^+ \cong \epsilon_0^r$ $(M-D)^+$, i.e. $T_H/T_D \cong 1$.

The fact that measured T_H/T_D ratios are usually slightly less than unity shows that the zero-point energy differences in the products are the more important. Reactions in which the ratio of T_H/T_D is considerably less than unity might be due to tunneling, but values considerably greater than unity are incompatible with a simple one-step $H^\cdot$ loss mechanism. Thus, in the case of triphenylphosphine oxide, where $H^\cdot$ loss occurs[275] by the reaction

and is necessarily accompanied by C—C and O—H bond formation, agreement with the above model would be fortuitous and is not found. The d_{10}-triphenylphosphine oxide having two fully deuteriated rings gives a large T_H/T_D ratio probably due to the effects of the new O—H or O—D bond upon the zero-point energies of products. Thus, $H^\cdot$ loss occurs with replacement of a C—D bond by a C—C and an O—D bond, while $D^\cdot$ loss can involve replacement of a C—H bond by a C—C and an O—H bond.

The above analysis does not cover cases where T is largely due to the non-fixed energy of the activated complex. In such cases, the ratio of T_H/T_D will depend on the internal energies of the metastable ions. This, as we have already seen, will be dependent upon the rate of increase of k with ϵ. For simple cleavages, $\epsilon^{\ddagger}$ will be a maximum and, therefore, T_H/T_D will be closest to unity. It is especially important that a single molecular ion with its single internal energy distribution be employed to determine T_H/T_D in these cases.

REACTIONS OF SMALL METASTABLE IONS

The bulk of the discussion of unimolecular fragmentation of metastable ions in this book has dealt with ions made up of more than
just a few atoms. The success of the quasi-equilibrium theory in
accounting for delayed fragmentations in these systems in terms of
the accumulation of internal energy in those modes appropriate to
pass over some potential energy barrier has been noted. There exists,
however, an entirely distinct set of processes which can lead to delayed unimolecular fragmentation. These involve forbidden predissociation and, therefore, slow fragmentation. Little is known about
the possible occurrence of such processes in large molecules, but
their occurrence in small ions has been documented in several cases.
The reaction of metastable ions of $H_2S^{+\cdot}$, shown below, falls into
the above class.

$$H_2S^{+\cdot} \rightarrow S^{+\cdot} + H_2$$

The vibrationally excited doublet ground electronic state of $H_2S^{+\cdot}$ is
metastable to emission (life-times for [ir] emission are $\sim 10^{-2}$ sec)
and it undergoes spin-forbidden predissociation to give the sulfur ion
in the quartet ground state. This example is discussed in more detail
below. Another case of spin-forbidden predissociation is provided by
the molecular ion of isocyanic acid which fragments to yield HCO^+
and atomic nitrogen with considerable (0.53 eV) kinetic energy
release[237].

$$HNCO^{+\cdot} \, (^2A'') \rightarrow HNCO^{+\cdot} \, (^4A'') \rightarrow HCO^+ \, (^1\Sigma^+) + N^\cdot \, (^4S)$$

It has been suggested[243] that the loss of N_2 from sym-triazole is
symmetry forbidden and for this reason sufficiently slow to be observed by means of a metastable peak.

A third predissociation process, rotational predissociation, has
been implicated in the slow fragmentation of small ions. For
example, HeH^+ ions undergo unimolecular fragmentation in a reaction chamber some 10^{-5} sec after ion formation[144, 239]. Proof that
a unimolecular process is indeed involved comes from the fact that
the angular distribution of the ionic product, H^+, was observed to be
isotropic in the center of mass. The reaction is suggested to occur by
tunneling of HeH^+ ions through the rotational potential barrier. Other reactions of diatomic ions have been shown to be due to spin-
forbidden predissociation. The reaction $N_2^{+\cdot} \rightarrow N^+ + N^\cdot$ is of this
type[218, 268].

Collision-induced reactions of small ions can be made to occur readily and a study of both unimolecular and collision-induced reactions can be combined to yield valuable thermochemical information. (The subject of collision-induced reactions is treated in detail later in this chapter.) The following discussion shows the value of the IKES technique in the study of small ions as exemplified by the case of H_2S. Its behavior provides an excellent illustration of how data from metastable ions can complement those obtained in other ways, particularly by photoionization appearance potentials and photoelectron spectroscopy. The combined information enables one to obtain a detailed understanding of the mechanisms of reactions of small ions. The improved thermochemical data which can result from such experiments are particularly important. The measurements of interest are the kinetic energy, Q', lost by the reactant ion in collision-induced reactions and the kinetic energy, T, released in the fragmentation step, whether unimolecular or collision-induced.

The formation of $S^{+\cdot}$ and H_2 from $H_2S^{+\cdot}$ occurs both unimolecularly and by a collision-induced process. Considering first the unimolecular reaction, the potential energy surface correlations of Fiquet-Fayard and Guyon[110] shown in Fig. 63 indicate that all three bonding states of $H_2S^{+\cdot}$ ($^2B_1, ^2A_1$ and $^2B_2$) might undergo predissociation via the repulsive $^4A_2$ state to form ground-state products, $S^{+\cdot}$ (^4S_u) and H_2 ($^1\Sigma_g^+$). The formation of ground-state products is

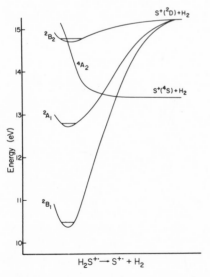

Fig. 63. Potential energy curves for $H_2S^{+\cdot} \rightarrow S^{+\cdot} + H_2$ illustrating possible predissociation of the $^2B_1$, $^2A_1$ and $^2B_2$ bonding states to give ground-state products.

required by the fact that the heat of formation of the first excited state of the products is too high to allow these states to be reached with the energy available in view of the value of the appearance potential. The appearance potential of $S^{+\cdot}$ formed by reaction of metastable ions is slightly lower (13.36 eV) than that for ions formed in the source (13.40 eV). Now, the photoelectron spectrum[97, 116, 224] of H_2S indicates a very low probability for population of states other than $^2A_1$ in the vicinity of the appearance potential of $S^{+\cdot}$. This state has vibrational structure implying that any predissociation is not rapid relative to the vibrational periods of the ion. However, predissociation could still be fast on the mass spectrometer time scale, and the reaction

$$H_2S^{+\cdot} \, (^2A_1) \rightarrow H_2S^{+\cdot} \, (^4A_2) \rightarrow S^{+\cdot} \, (^4S_u) + H_2 \, (^1\Sigma_g^+)$$

apparently provides the major mechanism whereby normal $S^{+\cdot}$ daughter ions are formed from $H_2S^{+\cdot}$. At energies well above threshold, there will also be contributions from the predissociation of the second excited ($^2B_2$) state, as well as direct excitation to the repulsive $^4A_2$ state. Predissociation of highly vibrationally excited ground state ions probably makes little contribution to product formation because of the low Franck—Condon factors for forming such ions with energies in the vicinity of the appearance potential of $S^{+\cdot}$.

The mechanism of fragmentation of metastable $H_2S^{+\cdot}$ ions can be interpreted in the light of the kinetic energy released in its reactions and in those of its deuteriated analogs. These results are summarized in Table 2 (ref. 157).

TABLE 2

Energy release in H_2S fragmentation

Reactant	Product	Kinetic energy release (eV)	
		Average	Maximum
$H_2S^{+\cdot}$	$S^{+\cdot} + H_2$	0.037	0.13
$HDS^{+\cdot}$	$S^{+\cdot} + HD$	0.075	0.17
$D_2S^{+\cdot}$	$S^{+\cdot} + D_2$	0.15	0.22

The kinetic energy release listed as average is the value measured from the width of the metastable peak at half-height, while the maximum release is determined from the width at the base line. This procedure is probably sufficiently accurate to enable good estimates to be obtained, although, strictly, the metastable peak should be deconvoluted to determine the range of kinetic energies released. Deconvolution procedures are presently being developed. It is significant that the range of energies involved in these reactions is rather small, even for $H_2S^{+\cdot}$ itself. This is seen qualitatively in the peak shape shown in Fig. 64 which should be compared with the metastable peaks typical of larger molecules, the sides of which generally have far shallower slopes (see Figs. 15, 44 and 61). The small range of

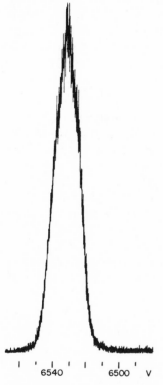

6540   6500   V

Fig. 64. Metastable peak for $H_2S^{+\cdot} \rightarrow S^{+\cdot} + H_2$, obtained by the HV scan technique at an analyser pressure of 3×10^{-5} Pa ($\sim 2 \times 10^{-7}$ torr).

translational energies observed in each case means that only a single vibrational level can be populated in the neutral hydrogen molecule. It is also noteworthy that the range of energies released decreases in the sequence $H_2S^{+\cdot} > HDS^{+\cdot} > D_2S^{+\cdot}$. The potential energy available for partition is slightly different for the three systems because of the different zero-point energies of H_2, HD and D_2. If the various hydrogen molecules are formed in their ground vibrational states, the kinetic energy releases can be corrected for the zero-point energy differences. When this is done, the corrected values of the maximum energy releases obtained are approximately constant down the series. This indicates that the products H_2, HD and D_2 are indeed formed in the zeroth vibrational state.

It is also concluded from these results that $S^{+\cdot}$ is formed by tunneling from the $^2A_1$ to the $^4A_2$ surface and that the maximum kinetic energy release corresponds to the difference in potential energy between the classical crossover point and the energy of the separated products. Ions having energy in excess of that required to surmount the activation barrier will predissociate rapidly (on the time scale of the mass spectrometer) and will comprise many of the normal daughter ions that have an appearance potential of 13.40 eV. The metastable ions tunnel through the barrier at a range of points, but the difference between the average and the maximum kinetic energy releases (0.09 eV) is consistent with an approximate difference of 0.04 eV between the appearance potential of daughter and that of metastable ions. The lower effectiveness of deuterium compared with hydrogen in tunneling through the barrier accounts for the much smaller range of kinetic energy releases in the reaction of metastable $D_2S^{+\cdot}$ ions since tunneling occurs in an even smaller energy interval immediately below the classical crossover point.

The foregoing interpretation suggests that all the energy available for partitioning is released as kinetic energy. Hence, the correction of the measured appearance potential of $S^{+\cdot}$ for the excess energy of the reaction is easily made. Indeed, this correction is very small in this particular case and an accurate value of $\Delta H_f^{\circ}(S^{+\cdot})$ is readily obtained.

The collision-induced reaction of H_2S to give $S^{+\cdot}$ occurs via different potential energy surfaces. If the signal obtained in the presence of collision gas is deconvoluted to remove the metastable peak, then the resultant signal has a width which corresponds to the release of 0.40 eV and its position is shifted to higher energy by 4.4 ± 0.5 eV (ref. 157). This is illustrated in Fig. 65. These values are independent of the nature or pressure of the target gas.

Most $H_2S^{+\cdot}$ ions reaching the collision region are in the lowest vibrational levels of the ground electronic state and a vertical excitation of 4.4 eV will give ions with energies in the vicinity of the second excited state. Predissociation of ions in this state via the repulsive $^4A_2$ surface seems likely to be the mechanism of fragmentation. Of the ~1.5 eV of energy available for partitioning in this reaction, only 0.4 eV, on the average, appears as translational energy. Hence, the most probable process involves the transfer of 1.1 ±

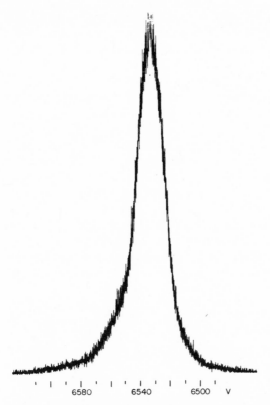

Fig. 65. Collision-induced and metastable peaks for the reaction $H_2S^{+\cdot} \rightarrow S^{+\cdot} + H_2$ obtained by the HV scan technique at an analyser pressure of 1.3×10^{-3} Pa (1×10^{-5} torr). The collision-induced component is recorded at higher energy. Note the break at approximately 6573 V.

0.5 eV into internal energy of the products. Because of the symmetrical nature of the fragmentation reaction, rotational excitation is considered to be small and, indeed, population of the vibrational state of H_2 with $\nu=2$, does correspond to an excitation of ∿1.1 eV. These points are all illustrated in Fig. 66. The maximum energy release as estimated from the base width of the signal is ∿0.7 eV, indicating that predominantly one level is populated. Indeed, there is evidence for a break in the collision-induced IKE peak shape (Fig. 65) corresponding to formation of the product H_2 as $\nu=1$.

The observation that, on average, only approximately 30% of the available energy appears as translational energy in this reaction (in contrast with 100% in the unimolecular reaction) is explained by the fact that there is sufficient energy to populate only the first vibrational level of H_2 in the unimolecular process.

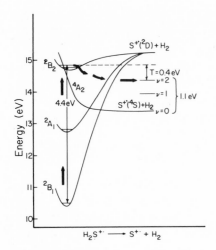

Fig. 66. Collisional excitation of $H_2S^{+\cdot}$ to yield ions in the $^2B_2$ state which predissociate via the $^4A_2$ state to yield ground-state $S^{+\cdot}$ ions and vibrationally excited H_2.

Before leaving the subject of unimolecular reactions in small ions, it is worth pointing out one more interesting observation. The molecular ion of NO^+ fragments, unimolecularly, to give $O^{+\cdot}$. When the metastable peak for this transition is examined under conditions of high energy resolution, fine structure is seen[119]. In the light of results from photoelectron spectroscopy, it has been shown that predissociation from two different electronic states is involved. The fine structure corresponds to reaction of ions in a number of vibrational levels of each electronic state.

COLLISION-INDUCED CHARGE TRANSFER REACTIONS

Internal energy can be given to a positive ion having a kinetic energy of the order of several thousand electron volts by passing it through a collision chamber containing a gas at low pressure. Depending upon how near the ion passes to a nucleus of the collision gas, various physical effects can be produced; the ion may be scattered out of the beam, the collision gas may be scattered or some of the kinetic energy of the ion may be converted into internal, electronic or vibrational energy of the ion or of the collision gas. A variety of reactions can readily be observed when a collision gas at a pressure of the order of 10^{-3} Pa ($\sim 10^{-5}$ torr) is introduced into the field-free region in front of the electric sector in a double-focusing mass spectrometer such as the Hitachi RMH-2. These reactions can also give rise to peaks in a mass spectrum if they occur after the electric sector, in the field-free region between it and the magnetic analyser. An electric sector measures the ratio of kinetic energy to charge in an ion beam and can readily be used to detect reactions in which the number, or sign, of the charges carried by the ion is changed. For example, doubly charged ions, in passing through the collision gas may charge exchange according to the reaction

$$m^{2+} + N \rightarrow m^{+} + N^{+}$$

where N represents a molecule of collision gas. If the high kinetic energy ions are not appreciably slowed down or deflected in the reaction, then their energy-to-charge ratio is approximately doubled and the product ions m^{+} will be transmitted if the voltage across the electric sector is changed from its normal value E to about $2E$. On the other hand, if a beam of positively charged ions is fired into the collision gas, ions of the same mass carrying two and three charges can be produced from singly and doubly charged reactant ions, respectively. These ions can be transmitted through the electric sector at approximate electric sector voltages of $E/2$ and $2E/3$, respectively. The net result of such reactions can be written

$$m^{+} \rightarrow m^{2+} + e$$

and

$$m^{2+} \rightarrow m^{3+} + e$$

Various amounts of internal energy can simultaneously be transferred to the collision gas causing it to acquire excitation energy, even sufficient to ionize it. If the voltage across the electric sector is reversed, ions can again be transmitted, showing that some negative ions having approximately the same kinetic energy as the original beam are being formed according to the reaction

$$m^{+} + 2e \rightarrow m^{-}$$

In this reaction, of course, the electrons can only be acquired from the collision gas so that one doubly charged positive ion or two singly charged positive ions are formed simultaneously.

For even longer-range reactions of the ions and collision gas molecules, smaller amounts of internal energy of the order of an electron volt can be transferred. These will be insufficient to change the degree of ionization of the ions or to ionize the collision gas, but they are sufficient to increase the rate of ion decomposition to such an extent that strong collision-induced peaks, similar in appearance to metastable peaks, can be observed. This aspect is discussed further in a later section of this chapter and in Chapter 5.

Symbolism used in describing these reactions

The reaction $m^{2+} \rightarrow m^+$ shown above is generally referred to as a charge-exchange reaction and the reverse reactions, such as $m^+ \rightarrow m^{2+}$, as charge-stripping reactions, but much more specific descriptions are necessary. Charge exchange and charge stripping may involve reaction or formation of ions carrying more than two charges and it is then necessary to specify how many charges are being transferred. In some cases when a singly charged ion passes through a collision gas and becomes further ionized, the collision gas itself acquires a positive charge.

A symbolism due to Hasted[131, 222] gives the number of charges carried by the reactants and also by the products. It uses the * symbol to indicate internal excitation. A few examples listed in Table 3 will serve to illustrate the system. The symbol N is used for the neutral gas molecule.

TABLE 3

Symbolism used to describe high energy ion—molecule reactions

Reaction			Symbolism
$m^+ + N$	$\rightarrow$	$m + N^+$	$10/01$
$m^+ + N$	$\rightarrow$	$m^{2+} + N^* + e$	$10/20^*$
$m^+ + N$	$\rightarrow$	$m^{2+} + N^+ + 2e$	$10/21$
$m^+ + N$	$\rightarrow$	$m^+ + N^+ + e$	$10/11$
$m^+ + N$	$\rightarrow$	$m^+ + N^*$	$10/10^*$
$m^+ + N$	$\rightarrow$	$m^- + N^{2+}$	$10/\bar{1}2$

Quantitative aspects of energy transfer

A good electric sector, such as that in the AEI MS9 or Hitachi RMH-2 mass spectrometers, capable of separating ion beams differing in kinetic energy by only one part in several thousand, is ideal for studying reactions such as those discussed above. The interaction of the ion and neutral molecule is complex and almost nothing is known about the detailed mechanisms by which energy is transferred. The interactions will usually involve transfer of linear and angular momentum as well as internal energy according to the laws of conservation of energy and momentum; there is a wide range of possibilities depending upon the direction and velocity with which the collision gas is scattered. The only source of energy for producing scattering and internal excitation is the energy of the ion. The collision gas will contain little internal energy; a stable or metastable ion that has reached the field-free region without fragmentation is unlikely to contain more than a fraction of an electron volt of vibrational energy and by far the greatest potential source of energy is the kinetic energy with which the ion is moving along the flight tube. The kinetic energy possessed by an ion moving along the central path through the mass spectrometer can readily be measured to a fraction of an electron volt by determining the voltage across the plates of the electric sector necessary to cause it to be transmitted. Thus, any energy lost by the ion in interacting with the collision gas can be measured to this same order of accuracy, provided that the ion is not appreciably deflected from its original direction of motion by the interaction.

It can be seen intuitively that the minimum effect upon the velocity and direction of the ion in the collision, when a particular amount of internal energy is given to the reactants, is produced when the collision gas (and the ion) are not scattered through appreciable angles. Scattering produces equal and opposite components of momentum at right angles to the original direction of motion of the beam; it tends to increase the kinetic energy given to the collision gas and thus to lead to a further decrease in the kinetic energy of the ion itself. The relationship between the angles through which ion and neutral gas molecules are scattered has been discussed[106] but, for simplicity, we shall only present equations for the case in which there is no appreciable scattering and we shall also neglect any components of angular momentum associated with the collision. Hopefully, the equations will still be useful in giving us information concerning the minimum energy transferred in such collisions. Indeed, as will be seen below, for collisions involving ionization, the minimum energy loss measured does correlate with ionization potentials and seems to justify the simplifications made in the following treatment.

Let the bombarding ions be of mass m_1 and let their velocities along the central path through the field-free region be v_1 before and v_3 after the collision. Let the neutral gas molecules be of mass m_2. We shall consider them to be, initially, at rest and to acquire, during the collision, a velocity v_2 in the direction of ion motion. As will be shown later, any velocity acquired by the collision gas in the direction perpendicular to ion motion will not affect the conclusions which follow. Let the collision result in a total internal energy gain by the products of Q. By the law of conservation of momentum

$$m_1 v_1 = m_1 v_3 + m_2 v_2$$

or

$$v_2 = \frac{m_1 (v_1 - v_3)}{m_2} \tag{40}$$

By the law of conservation of energy

$$\tfrac{1}{2} m_1 v_1^2 = \tfrac{1}{2} m_1 v_3^2 + \tfrac{1}{2} m_2 v_2^2 + Q$$

or

$$m_2 v_2^2 = m_1 v_1^2 - m_1 v_3^2 - 2Q \tag{41}$$

Substituting (40) in (41) we have

$$\frac{m_1^2 (v_1^2 - 2v_1 v_3 + v_3^2)}{m_2} = m_1 v_1^2 - m_1 v_3^2 - 2Q$$

Collecting terms

$$v_3^2 \, m_1 (m_1 + m_2) - v_3 \, 2m_1^2 v_1 + [m_1(m_1 - m_2)v_1^2 + 2Qm_2] = 0$$

$$\therefore v_3 = \frac{m_1^2 v_1 \pm \{m_1^4 v_1^2 - m_1(m_1 + m_2) [m_1(m_1 - m_2)v_1^2 + 2Qm_2]\}^{1/2}}{m_1(m_1 + m_2)}$$

$$= \frac{v_1}{m_1 + m_2} \, [m_1 \pm m_2 \left(1 - \frac{2Q(m_1 + m_2)}{m_1 v_1^2 m_2}\right)^{1/2}]$$

$$= \frac{v_1}{m_1 + m_2} \, [m_1 \pm m_2 \left(1 - \frac{m_1 + m_2}{m_2} \frac{Q}{eV}\right)^{1/2}]$$

where eV is the kinetic energy $\tfrac{1}{2} m_1 v_1^2$. Ignoring the negative sign before the square root

$$v_3^2 = \frac{v_1^2}{(m_1 + m_2)^2} \, [m_1^2 + m_2^2 \left(1 - \frac{Q}{eV} \frac{m_1 + m_2}{m_2}\right)$$

$$+ 2m_1 m_2 \left(1 - \frac{Q}{eV} \frac{m_1 + m_2}{m_2}\right)^{1/2}]$$

$$= \frac{v_1{}^2}{(m_1 + m_2)^2} \left[m_1{}^2 + m_2{}^2 - \frac{Q}{eV} (m_1 + m_2) m_2 \right.$$

$$+ 2m_1 m_2 \left\{ 1 - \frac{Q}{eV} \frac{m_1 + m_2}{2m_2} - \frac{(m_1 + m_2)^2}{8m_2{}^2} \left(\frac{Q}{eV} \right)^2 - \ldots \right\}]$$

$$\cong \frac{v_1{}^2}{(m_1 + m_2)^2} [(m_1 + m_2)^2 - \frac{Q}{eV} (m_1 + m_2)^2 -$$

$$(m_1 + m_2)^2 \frac{m_1}{4m_2} \left(\frac{Q}{eV} \right)^2] = v_1{}^2 [1 - \frac{Q}{eV} - \frac{m_1}{4m_2} \left(\frac{Q}{eV} \right)^2]$$

Now

kinetic energy given to m_2 = $\frac{1}{2} m_2 v_2{}^2$

$$= \frac{1}{2} m_1 v_1{}^2 - \frac{1}{2} m_1 v_3{}^2 - Q$$

$$= \frac{1}{2} m_1 v_1{}^2 - \frac{1}{2} m_1 v_1{}^2 [1 - \frac{Q}{eV} - \frac{m_1}{4m_2} \left(\frac{Q}{eV} \right)^2] - Q$$

$$= Q + \frac{m_1}{4m_2} \frac{Q^2}{eV} - Q$$

$$= \frac{m_1}{4m_2} \frac{Q^2}{eV} \tag{42}$$

Thus, the ratio of the kinetic energy of m_2 to the initial kinetic energy of $m_1{}^+$ is given by

$$\frac{m_1}{4m_2} \left(\frac{Q}{eV} \right)^2$$

If Q is of the order of 10 eV and the kinetic energy of the ions, $m_1{}^+$, is of the order of 10^4 eV, the above quantity is negligible compared with the difference $\frac{1}{2} m_1 v_1{}^2 - \frac{1}{2} m_1 v_3{}^2$. Thus, when the minimum kinetic energy is taken from the ion in order to produce excitation without scattering, the loss of ion kinetic energy (Q') is a direct measure of the energy of excitation. This extremely valuable simple relationship comes about because of the high kinetic energy of the ion. Equation (42) shows that the ratio in which kinetic energy change is shared between m_1 and m_2 is given by

$$Q \Big/ \left(\frac{m_1 Q^2}{4m_2 \, eV} \right) = \frac{4m_2 eV}{m_1 Q}$$

When $Q = 10$ eV and the ion energy is 10^4 eV, this ratio is 4×10^3 when $m_2 = m_1$. In such a case, the error in neglecting the kinetic energy given to m_2 is negligible for the process involving no scattering. The favorable "disproportionation factor" that operates as far as partitioning of the kinetic energy change between m_1^+ and m_2 is concerned is a direct consequence of the high initial energy of the ion. The factor depends upon the ratio of the masses m_1 and m_2. The accuracy in neglecting the energy of m_2 is greatest when $m_1 \gg m_2$. To take an extreme case, for ions of xenon ($m_1 = 132$) impinging on helium collision gas ($m_2 = 4$) the ratio in which energy is shared becomes 1.2×10^2. Even in this case, a correction only of the order of 1% must be applied to the measured value of Q', the kinetic energy change of m_1^+, to give the internal energy Q.

Measurement techniques employed

The conditions for collection and determination of the mass-to-charge ratio of the product ions of collision-induced reactions, especially those in which the number of charges carried by the reactant ions is changed, need to be considered carefully. It is worthwhile to derive the general relationship that will enable the product ions to be measured (at the final collector of a double-focusing mass spectrometer of the type in which the electric sector precedes the magnetic sector) by scanning the accelerating voltage at fixed electric sector voltage. Suppose that the reactant ions have a mass m_1 and carry x charges, and that it is desired to record product ions of mass m_2 and carrying y charges formed in a collision-induced transformation. Suppose that the magnetic sector is set to transmit ions of mass-to-charge ratio m_z at the normal values V and E of the accelerating and electric sector voltages, respectively. (This value, m_z, can conveniently be thought of as the reading given by a mass-marking device measuring the magnetic field.) If the values of V and E are switched, together, to some multiple, c, times their initial values, the actual mass-to-charge ratio of the ions transmitted by the magnet will change; the mass scale will vary inversely with the energy of the ions entering the magnetic sector (see eqn. (5), p. 11) and the mass value indicated by the mass marker must be multiplied by $1/c$ to give the true mass-to-charge ratio of the ions recorded. For cases in which fragmentation or alteration of the number of charges carried by an ion occurs in the first field-free region, the value of V is relevant only to the energies of the reactant ions and it is the value of the electric sector voltage only that determines the energy-to-charge ratio of the product ions selected for transmission into the magnet. In the case under consideration here, to transmit the product ions the electric sector voltage must be adjusted from E to E_1 where

$$E_1 = \frac{m_2 x}{m_1 y} \, E \qquad (43)$$

The magnet will transmit ions of mass-to-charge ratio $m_z\{m_1 y/(m_2 x)\}$ at this setting of the electric sector, and if the product ions of mass-to-charge ratio m_2/y are to be collected, we must have

$$m_z \frac{m_1 y}{m_2 x} = \frac{m_2}{y}$$

or

$$m_z = \frac{m_2^2 x}{m_1 y^2} \qquad (44)$$

The magnet must thus be set to the mass value at which a metastable peak would be observed if the reaction were unimolecular and occurred in the second field-free region.

Let us consider as examples of the operation of eqns. (43) and (44), the settings necessary to transmit the product ions of the charge permutation reactions given at the beginning of this section. In charge-exchange reactions 20/10, the electric sector voltage must be changed to $2E$ and the magnet mass marker set at a value of $2m$; in charge-stripping reactions 10/20, the electric sector is set at $E/2$ and the magnet at $m/4$; in charge-stripping reactions 20/30, the electric sector is set at $2E/3$ and the magnet at $2m/9$.

Production of multiply charged ions

One of the phenomena observed when a high-energy beam of singly charged ions is fired into a collision region containing neutral gas atoms or molecules at low pressure is the formation of ions carrying two or more charges according to reactions such as 10/20 and 10/30. Let us consider a typical result obtained when the electric sector voltage is scanned over the region around $E/2$ so as to detect ions carrying two positive charges that have been formed in the collision region from singly charged ions. Figure 67 (a) shows the signal obtained when argon ions carrying a single positive charge are fired into air used as collision gas. Three peaks are observed. Peaks B and C correspond to conversion of two different amounts of kinetic energy into internal energy; peak A is detected exactly at the sector voltage $E/2$ and corresponds, therefore, to doubly charged ions formed without loss of any ion kinetic energy. The difference, Q' between the

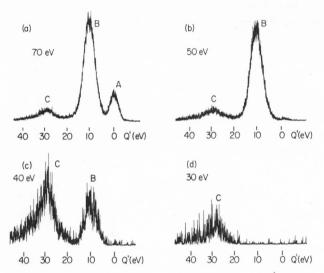

Fig. 67. Kinetic energy loss spectra for the reaction $Ar^{+\cdot} \rightarrow Ar^{2+}$ using air as collision gas. The effect of varying the electron energy upon the relative abundances of peaks A, B and C is shown.

kinetic energy of the reactant and product ions is indicated for each peak in the figure.

Similar spectra can be obtained from more complicated molecules. Studies on nitric oxide[88], for example, have shown the existence of two peaks both corresponding to the formation of $NO^{2+\cdot}$ ions but with transfer of different amounts of kinetic energy into internal energy. Most previous studies of 10/20 reactions have been concerned with rare gases, but both Jennings[154] and Seibl[240], although severely hampered by the very weak signals they could attain, have detected products which would correspond to this reaction from polyatomic organic ions. Studies using the IKE method have also shown that peaks corresponding to an electric sector voltage of $E/2$ can readily be obtained from organic molecules. A peak at this value of the electric sector voltage could arise from the unimolecular or collision-induced decomposition of ions carrying a single positive charge into two fragments of equal mass as well as by the 10/20 reaction path. Peaks due to the former reaction are always broadened by kinetic energy release; the peaks due to the latter process are always much narrower and the two processes can be distinguished on this basis alone. Consider the reaction

$$m^+ \rightarrow \tfrac{1}{2}m^+ + \tfrac{1}{2}m$$

occurring in the first field-free region. The product ions will be transmitted through the electric sector and will be recorded by the mass marker in the magnetic field as having an apparent mass $m/4$. They will thus overlie the peaks due to the 10/20 reaction of m^+ ions. The method of recording is to vary the accelerating voltage over a small range. The peak due to the fragmentation reaction, if it is due to a unimolecular decomposition, will appear at exactly the value V. Even if it is collision-induced, it will appear very close to this value. It may be accompanied by other peaks in which ions of mass-to-charge ratio $m/2$ are formed from different parent ions. For example, if ions of mass $(m + 1)$ or $(m - 1)$ also decompose to form $\frac{1}{2}m^+$ ions, these will be transmitted through the electric sector and recorded by the magnet as the accelerating voltage is scanned over a wider range. Figure 68 gives an example of what may occur in practice. It shows an HV scan in benzonitrile with the magnet mass marker set to transmit ions of apparent mass 19. A value of the accelerating voltage of 7536 V corresponded to twice the value required to transmit the main beam. A sharp peak, centered at about 7552 V and due to the transition $76^+ \rightarrow 76^{2+}$ in benzonitrile can be seen. A continuum of overlapping broad peaks can also be distinguished centered at 7635, 7536, 7438 and

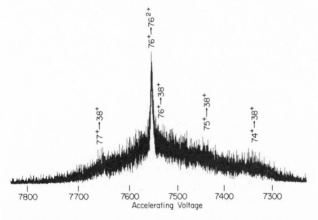

Fig. 68. Peaks of m/e 38 arising from charge stripping (the narrow peak at 7552 V) and fragmentation reactions in the field-free region. The ions were collected by scanning the accelerating voltage over a small range in the vicinity of twice the value needed to transmit the main ion beam.

7339 V due, respectively, to the fragmentation of ions having a single charge and of masses 77, 76, 75 and 74 to give ions of mass-to-charge ratio 38. It may be noted, in passing, that the sharp peak shows an energy loss of about 16 eV (more accurately

15.7 eV). It will be shown later that this can be correlated with the
difference between the double and single ionization potentials of
C_6H_4. In some other cases, the relative heights of the sharp peak of
interest and the broad peaks is such that accurate measurement of
the position of the sharp peak becomes impossible. Fig. 69(a) shows
a plot obtained for benzene at an apparent mass of 19.5 (*i.e.* 78/4).
The reactions $78^+ \rightarrow 78^{2+}$ and $78^+ \rightarrow 39^+$ both contribute to the
peak seen, but the signal from the latter reaction is much stronger
than that from the former. In such cases, it is usual to study, instead,
the peak due to ions containing a single heavy isotope and so the
magnet is set to an apparent mass 19.75 at which the product ions of
the reaction $79^+ \rightarrow 79^{2+}$ are observed. An HV scan of this reaction is
shown in Fig. 69(b). It can be seen that there is now no interference
from decomposing singly charged ions.

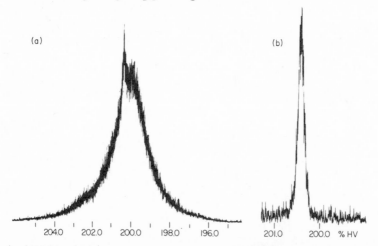

Fig. 69. HV scans with the ratio of electric sector to high voltage settings
changed to half the normal value. The curves refer to benzene. (a) Peaks at
apparent m/e 19.5 arising from charge stripping and from the fragmentation
reaction $78^+ \rightarrow 39^+ + 39$. The fragmentation reaction is accompanied by energy
release and gives rise to the broad peak. (b) Peak at apparent m/e 19.75 showing
the charge-stripping peak arising from molecular ions containing a heavy isotope.

It is also possible to plot a mass spectrum by setting the accelerating voltage and electric sector voltage to the values V and $E/2$ and
then scanning the magnetic field. The separation in energy between
the sharp and broad peak centers is not large (in the example shown
in Fig. 69 it corresponds to only about 1 part in 500) and if the β-slit
is widened, the ability to resolve the two peaks will be lost. In the
Hitachi RMH-2, for example, a change of 1 part in 500 in energy

corresponds to a displacement in the β-slit plane of about 1mm, so for wide β-slit settings it is possible to scan a mass spectrum in which peaks due to both processes can be seen. Figure 70 shows such a spectrum in the case of aniline and at some mass numbers, for example, 28, 50, 52 and 54, a broad peak can be seen beneath the narrower peak due to the 10/20 process. It is worth mentioning that

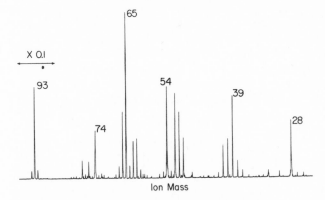

Fig. 70. $E/2$ mass spectrum of aniline.

the intensity of the "$E/2$ spectrum" that can be obtained in this way is sufficiently great for it to be used as a means of identifying aniline in the same way as the normal mass spectrum. Little is known about the structural features that might be discernible from such a spectrum because so few of them have been plotted, but they provide a complex, and thus potentially useful, alternative to conventional spectra. If the β-slit is narrowed, then changes will be observed in the spectrum due to the fact that the two kinds of product ions will not be transmitted at a single value of the accelerating voltage. Figure 71 shows the effect of changing V by a discrete amount at a fixed electric sector voltage, $E/2$, and then scanning the magnetic field. It can be seen that as the accelerating voltage is increased, more of the peaks due to the 10/20 process are transmitted and that, conversely, at lower accelerating voltage, it is mainly the broad peaks that are observed. For example, very little of the sharp peak due to the 10/20 reaction of mass 51 appears in Fig. 71(b).

It is now necessary to look in greater detail at the peaks obtained from the rare gases in an attempt to understand more about the nature of the information being acquired. Early work, not using IKE methods, revealed that three distinct mechanisms can operate to form doubly charged ions from singly charged ions having kinetic

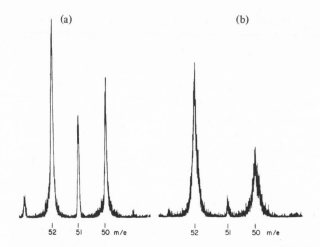

Fig. 71. Effect of changing the accelerating voltage on the transmission of ions resulting from charge-stripping (10/20) and fragmentation reactions while maintaining the electric sector at $E/2$. Part (a) was obtained at exactly V and (b) at 0.9974 V.

energies in the keV range. These are (i) autoionization of the singly charged ions[94, 217] from states with an energy in the immediate vicinity of the double ionization potential of the rare gas in question, (ii) a pseudo-unimolecular process involving interaction with a solid (slit) surface possibly from high Rydberg states[216] and (iii) formation of the doubly charged ions in a collision process involving a neutral gas molecule and also involving the higher Rydberg levels, at least in the case of argon[184]. All of these experiments were done with poor energy resolution and thus we would hope that the use of the IKE technique would extend our understanding of the processes involved and possibly lead to new and more accurate values for energy levels in highly excited and multiply charged ions.

The bombarding argon ions used in the experiment illustrated in Fig. 67(a) were produced using 70 eV ionizing electrons. As the energy of the electrons is reduced progressively to about 50 eV, 40 eV and 30 eV (see Fig. 67(b), (c) and (d)), peak A disappears, followed by peak B, until finally peak C becomes attenuated. From the positions of the peaks, it is estimated that the loss of kinetic energy, Q' from the $Ar^{+\cdot}$ ions in order to form Ar^{2+} ions is, for peak A, 0 ± 0.3 eV; for peak B, 8.9 ± 0.8 eV and for peak C, 27.5 ± 0.6 eV. Values of the minimum energy necessary to form Ar^{2+} and $Ar^{+\cdot}$ from neutral Ar are 43.4 and 15.9 eV, respectively. The difference between these values is 27.5 eV. This is, therefore, the energy necessary to form Ar^{2+} in its ground state from $Ar^{+\cdot}$ also in its ground state. Peak C would seem to be due to this process. It is noteworthy that peak C can still be observed even when 30 eV bombarding electrons are used to produce the $Ar^{+\cdot}$ ions. Of course, only 15.9 eV is necessary to produce the reactant $Ar^{+\cdot}$ ions in the ground state.

There are known doublet long-lived excited states of $Ar^{+\cdot}$, ($^2F_{7/2}$), requiring a mean minimum energy of 34.3 eV for their formation from ground-state Ar. These ions would require a further minimum energy of (43.4—34.3) = 9.1 eV to convert them to ions Ar^{2+}. Peak B would seem to correspond to this process. The peak cannot be observed when 30 eV electrons are used to produce the $Ar^{+\cdot}$ ions and is very small at 40 eV, confirming that the appearance potential of the excited reactant ions lies between these two energies. The doublet spacing is only a small fraction of an electron volt and the energy resolution used is insufficient to distinguish the two individual components. The exact position of the observed peak is discussed in greater detail below. Peak A seems to correspond to the formation of Ar^{2+} from $Ar^{+\cdot}$ ions without the transfer of any kinetic energy into internal energy. This may, therefore, be due to the collisional process involving high energy Rydberg states of $Ar^{+\cdot}$. Peak A could only be produced using bombarding electrons having an energy in excess of 43.4 eV and, indeed, from Fig. 67(b) it can be seen that the peak has almost disappeared when 50 eV electrons are employed.

It is thus seen to be possible to use experiments of this kind to measure the difference between single and double ionization potentials and to detect and measure the internal energies of long-lived excited states of ions. The widths of the peaks in the example shown can be seen to be of the order of ± 2.5 eV at half-height so that the position of a maximum can readily be determined to an accuracy of the order of 0.5 eV. The useful resolution is such that components of a doublet can be detected if they are separated in energy by the order of 1 or 2 eV. An example is provided by peak C in the case of xenon[12] which represents the transition of ground state $Xe^{+\cdot}$ ions into Xe^{2+} ions. The ground state of $Xe^{+\cdot}$ is a doublet designated $^2P_{1/2}$ and $^2P_{3/2}$. The energy difference between these states is 1.3 eV and seems to be just detectable as fine structure.

The relative abundances of the peaks A, B and C depend upon the pressure of collision gas as well as the energy of the bombarding electrons. At very low pressures, the abundance of peak A is enhanced relative to the other two. This reflects the fact that no kinetic energy needs to be lost in the collisions leading to peak A; these collisions have the greatest cross-section and can occur at the longest range interaction between the ion and neutral. As the collision gas pressure is increased, the chance is increased that an ion will pass within a closer distance to a collision gas molecule than the minimum necessary to produce reaction A, and the relative abundance of peaks B and C will increase. As the pressure is increased further, the abundance of C will continue to increase relative to A and B. The relative abundances observed also depend upon the kinetic energy of the bombarding ions. As the accelerating voltage is increased, the abundance of peak C increases relative to the other two. Either this interaction, involving the greatest transfer of energy, now takes place with less discrimination against the product ion due to scattering (the scattering angle being inversely dependent on total kinetic energy) or the cross-section for reaction C increases most rapidly with ion energy. Perhaps the most important single feature in the spectra is the shape of the observed peaks. The non-reacting ions form a beam having a finite spread of kinetic energy and this is broadened by elastic collisions with neutral gas molecules as it passes through the collision chamber. The resultant ion beam provides a standard for intrinsic beam width in the presence of collision gas. The actual kinetic energy of the ions m^{2+} formed from m^+ ions will be reduced, not only by the energy necessary to ionize m^+ but also by an amount corresponding to internal energy transferred into the m^{2+} ions or neutral molecules and to scattering. Referring to Fig. 72, it can be shown[12] that the relative velocities of the ions and collision gas

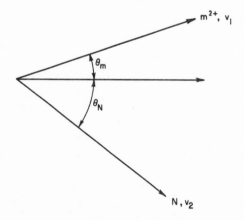

Fig. 72. Angular distribution and velocities of the products m^{2+} and N of the charge-stripping reaction 10/20.

molecules are connected by the relationship

$$\frac{v_2}{v_1} = \frac{m \sin \theta_m}{N \sin \theta_n}$$

where m and N, v_1 and v_2, θ_m and θ_n refer to the ion and neutral gas molecules, respectively. The value of θ_m, and hence the extent of discrimination by the instrument, is therefore dependent on the ratio of the masses m and N. The experimental results bear out this general conclusion. For example, Fig. 73 shows that when helium is used as collision gas with argon ions, peak C is broader than when air is used as collision gas (see Fig. 67). More of the scattered ions are recorded

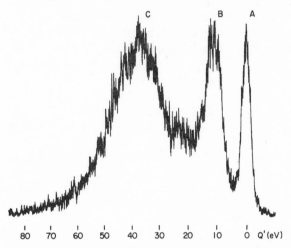

Fig. 73. Kinetic energy loss spectrum for the charge-stripping reaction (10/20) of argon ions impinging upon helium.

in the case of helium; the instrumental discriminations are least in this case. In order to obtain meaningful correlations of kinetic energy loss and the relevant energy levels, it is necessary to measure the kinetic energy value corresponding to the onset of the process where the effects due to scattering and transfer of more than the minimum energy are at their lowest value. Indeed, it is found that the slope of the peak side corresponding to onset is the steeper (see Fig. 70). Therefore, the procedure adopted in measuring the position of a peak is to determine the value at which onset is detected. When this is done, it has been found[12] that the values calculated for double ionization potentials in the rare gases and the values calculated for ionization from the long-lived excited states of the ion agree with the best literature values to within 1 eV.

Besides the three types of peak designated A, B and C that are commonly observed in the energy loss spectra of the rare gases, a fourth peak is detected when $Ar^{+\cdot}$ ions are fired into Ar collision gas. A typical spectrum is shown in Fig. 74 and the presence of a peak displaced by about 11.5 eV from the onset of peak C can be discerned. It is thought that this is an example of a 10/20* reaction and that the peak represents the formation of the doublet $^3P_0$ and $^3P_2$ excited states of neutral argon that are 11.6 eV above the ground state. A 10/21 reaction (in which the collision gas is also ionized) has been detected in the case of $Ne^{+\cdot}$ ions fired into Ne collision gas. Apart from isolated examples of this kind, in which extra peaks appear, the positions of the energy-loss peaks are unaffected by the collision gas used although the cross-sections for the various processes are altered considerably when the collision gas is changed. Even large organic molecules can be used as efficient target species.

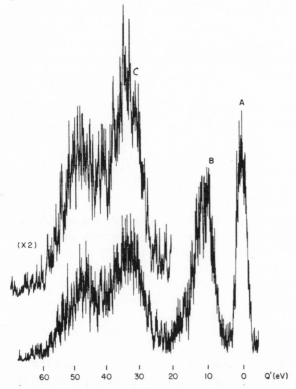

Fig. 74. Charge-stripping reaction (10/20) of argon ions impinging upon argon giving rise to peaks A, B and C. The extra peak with an onset at approximately 40 eV is due to the reaction 10/20* in which the collision gas is excited.

The same techniques can also be used in studying organic ions. Usually, the spectra obtained in such cases correspond to the presence of only a single peak (peak C in the nomenclature used with the rare gas ions) at each mass number. Examples have already been given for benzonitrile and benzene (see Figs. 68 and 69). The method can be applied in principle to any organic ion and is not necessarily restricted to molecular ions. It has the advantages of high sensitivity, of a favorable shape of efficiency curve as a function of kinetic energy loss (this should be compared with the parabola-shaped ionization efficiency curve for formation of doubly charged ions by electron bombardment), of the fact that an ion beam that is almost monoenergetic is available and that interference from peaks due to other processes do not often occur and can usually be overcome by the methods discussed above (see p. 138). The shape of the peak from an organic ion is strikingly different from the corresponding peak (peak C) seen from the rare gases, falling more sharply to zero at its trailing edge. The absence of a tail shows that the probability that the ion will receive appreciable excitation during the ionization without fragmentation is low and also suggests that the probability that the ion will be scattered through an appreciable angle without fragmentation is small. Literature values for double ionization potentials of organic compounds are available in some cases[114] and where these are known, a comparison has been made with the value obtained in the IKE method by adding the energy loss figure to the first ionization potential for the neutral organic molecule. In all of the six cases so far studied[90], agreement between experimental and literature values was better than 0.3 eV and although this may be fortuitously good, it suggests that the IKE method will indeed be capable of determining many new values of double ionization potentials. The method also seems capable, in theory, of measuring triple ionization potentials using either the reaction

$$m^{2+} + N \rightarrow m^{3+} + N + e$$

or

$$m^+ + N \rightarrow m^{3+} + N + 2e$$

In practice, only the first of these reactions was observed in the only case studied so far. A value of 41.2 ± 1 eV was found for the triple ionization potential of naphthalene, in agreement with the literature value of 40 ± 5 eV. Little is known concerning the details of the interaction of the ion and neutral molecule in a charge-stripping collision. It must be remembered that the velocity of a 10 keV ion of mass 100 is only about 3% of that of a 50 eV electron and the cross-section for collision, even if it involves transfer of some 20 eV of energy, is large so that a strong interaction over a time of at least 10^{-14} sec can be expected between the ion and neutral and weaker interaction will extend over longer times still. A C—H stretching vibration has a frequency of about 10^{14} sec^{-1} so that appreciable movement of nuclei can take place during the collision which will not, therefore, take place under strict Franck—Condon conditions. We have already seen that when double ionization by electron bombardment occurs, this is a truly "vertical" process and two electrons are removed from the molecule very rapidly. Here, by contrast, the removal of the second electron could tend more towards "adiabatic" conditions. The first electron will have been removed by electron bombardment in the ion chamber, but the resultant ion will have moved towards the field-free region over a period of several microseconds in which time the nuclei will have been able to move towards equilibrium conditions. This might make possible the formation of a doubly charged ion by charge stripping in a lower state of excitation than could be achieved by formation of doubly charged ions directly from neutral molecules in the source. Formation of lower-energy fragment ions carrying two charges might also be possible by charge stripping of singly charged fragments. Such ions could be distinguished by their longer lifetimes and do seem to have been detected in the case of pyridine. No doubly charged ions of mass 53 are detected in the normal mass spectrum nor in the $2E$ spectrum in which doubly charged ions entering the field-free region are converted into singly charged ions by charge exchange. However, they are observed in the $E/2$ spectrum and this means that in this case, they are sufficiently stable upon formation to travel to the final collector without decomposition.

Charge-exchange processes

The reaction 20/11 in which molecules of a collision gas transfer one of their electrons to a doubly charged ion have already been mentioned briefly. The mass spectrum of all the ions that result from such a reaction can be obtained by setting the electric sector voltage to twice its normal value and scanning the magnetic field. A typical $2E$ spectrum is shown in Fig. 75. Such spectra can typically be obtained at a sensitivity of the order of 10^{-3} of that for normal mass

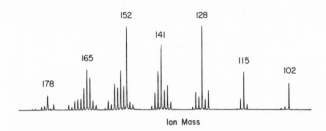

Fig. 75. 2E mass spectrum of 1,2,3,4,5,6,7,8-octahydroanthracene.

spectra and their use as "fingerprints" of organic molecules and in giving structural correlations[11] has been noted. This aspect is mentioned in Chapter 5 and the information that can be deduced from the relative abundances of the peaks observed concerning the stabilities of doubly charged ions is treated in Chapter 6. Here, we shall be mainly concerned with the information about energy levels that can be inferred from charge-exchange reactions. The mass values indicated by a mass marker measuring the sector magnetic field will correspond to twice the true masses when the electric sector is set to $2E$ (see eqn. (44) where $m_1 = m_2$, $x = 2$, $y = 1$). The peak shapes obtained are sharp at the low collision gas pressure normally used, $<10^{-3}$ Pa ($< \sim 10^{-5}$ torr); the interaction is a weak, long-range one and therefore not likely to cause appreciable deflection of the product ions. If too high a collision gas pressure is used, the observed peaks broaden and their height is reduced, as a result of scattering and consequent discriminatiom effects.

On their passage from the first field-free region to the final collector, the singly charged ions formed by charge exchange may decompose unimolecularly. The product ions from such decompositions can be detected by the HV scan technique or as metastable peaks in the mass spectrum in just the same way as can the products of decomposition of metastable ions formed in the source. It is often the case that doubly charged ions may be known to have a different structure from singly charged ions formed from neutral molecules. When these doubly charged ions are converted into singly charged ions by charge exchange, the product ions will have the structures of the parent doubly charged ions from which they have been formed. The kinetics and energetics of the fragmentation of such ions as determined from the abundance and width of the associated metastable peaks can be used to give structural information as discussed in Chapter 6.

As in the case of the 10/20 spectra that have been discussed at length above, fine structure can be seen in the kinetic energy spectra of the products of the charge-exchange reaction. A unique feature of these spectra is that the peak corresponding to the products of highest kinetic energy sometimes corresponds to *more* than twice the energy of the original ion beam. This is, at least in part, a consequence of the sharing of the two charges between the ion and the collision gas and arises in the following way. A doubly charged ion carries a large amount of potential energy by virtue of the mutual repulsion of the two charges. This energy is released as kinetic energy if the charges separate either in a fragmentation process or by charge transfer. In the case of unimolecular fragmentation, as already discussed in Chapter 3, the product ions can be considered as separating from each other (in a frame of reference moving with the center of mass) with components of velocity in any direction over a solid angle of 4π. It follows that in a fixed frame of reference, some of the ions will acquire an increased velocity along their original direction of motion while others will be slowed down or given an extra velocity laterally. This symmetry does not obtain in the case of charge transfer. Here, the doubly charged ions move in a single direction through molecules of collision gas that are essentially at rest. At some critical distance, not necessarily the minimum distance at which the ion approaches the neutral, an electron is transferred and the ion will continue its motion past the newly formed ion. Repulsion between the two charges will then lead to an increase in the velocity of the ion and give a peak in the ion kinetic energy spectrum at a position higher than would correspond to an electric sector voltage of $2E$. The fraction of the total available energy given to the ion will be large; a "disproportionation" factor like that operating in the case of charge stripping and discussed above will be involved. The range of directions that can be given to the ion by this repulsive force will depend upon the distance between ion and neutral when charge exchange occurs and two cases are illustrated diagrammatically in Fig. 76. In

Fig. 76. Energy gain accompanying the charge exchange reaction 20/11. In (a), charge exchange occurs at the point of closest approach of the two particles and the resulting ions repel one another as they separate. In (b), although electron transfer may occur at the same distance, the ions continue to approach each other after charge exchange.

case (a) which represents an extreme where charge transfer takes place exactly at the minimum distance of approach, the initial force on the ion is at right angles to its direction of motion but will act more nearly along the direction of motion as the charges separate. In case (b) the ion approaches closer to the ionized collision gas molecule after charge exchange. There is again a lateral component to the force, which reaches its maximum as the ion passes closest to the ionized collision gas. In either case, the fast-moving ion will be slightly deflected from its original direction and, in addition, given almost all of the energy e^2/R resulting from separation of the two charges. In case (b), less of the resultant velocity will be along the original direction of motion. An electric sector is capable of focusing a monoenergetic beam of ions that has a small angular spread, so provided the deflection given to the ion is not too great, its correct energy may be recorded by the electric sector. Complications will arise due to off-axis ions that are scattered back on to the central path during the collision and this effect will change the peak shape observed. The amount of energy given to the ion will provide a direct measure of the distance R at which charge exchange occurred; this assumes that in the charge exchange process, the ion does not provide any internal energy from its own translational energy. The shape of the curve produced in this "knock-on" process will give information about the range of values of R over which charge transfer occurs; the tail of the curve on the high-energy side will give information concerning the minimum value of R. A typical "energy gain" curve is shown in Fig. 77 which was obtained by firing helium doubly charged ions of kinetic energy 4 keV into krypton. The high-energy peak is broad, shows a maximum height corresponding to an energy gain of about 6.8 eV and a maximum energy 5 eV greater. There is a much stronger, narrow peak corresponding to an energy loss of 0.8 ± 0.5 eV. This is thought to be due to the conversion of He^{2+} in its ground state to He^{+*} in its lowest energy excited state, releasing an energy of 13.7 eV. It requires 14.3 eV to produce Kr^{+*} in its ground state leaving a calculated energy deficit of 0.6 eV, agreeing closely with the observed value for the kinetic energy loss by the helium ions. It is generally true that energy-loss peaks are most intense when the energy deficit is small (<1 eV) as in the present case. It should be noted that the energy balance in the charge-exchange reaction will contain a term to take account of the coulombic repulsion energy e^2/R. Generally, results obtained by ignoring any correction to the energy from this term match the observed energy losses within experimental error, suggesting that

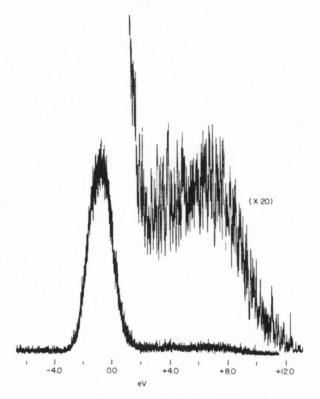

Fig. 77. Kinetic energy spectrum due to the charge-exchange reaction (20/11) for He^{2+} on Kr. A weak energy gain peak and an intense energy loss peak can be seen.

these are long-range reactions for which e^2/R may be as low as 0.5 eV, *i.e.* R could be as large as 30Å. In contradistinction, the energy-gain peaks correspond to very much closer approach of the ion and neutral before charge transfer takes place. The values of R corresponding to the mean and extreme energy gains of 6.8 and 11.8 eV, *viz.* 2.1 and 1.2Å, provide a measure of the closest distance of reaction. These values are actually minimum values since no account has been taken of the conservation of angular momentum in these close interactions.

Charge transfer will, presumably, take place by a curve-crossing mechanism; at some small radius the potential energy curve for the product state (m$^+$ + N$^+$) will cross with that for the reactant state (m^{2+} + N). The force between the reactants is largely of the van der Waals type and the crossing is illustrated schematically in Fig. 78.

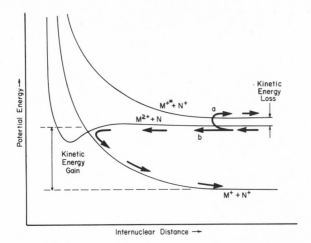

Fig. 78. Potential energy diagram illustrating the two mechanisms by which charge exchange (20/11) may occur. The long-range reaction is accompanied by a small kinetic energy loss, the short-range reaction by a considerable kinetic energy gain.

The force between the products is coulombic in the case under discussion but the phenomenon may well be more general. This figure also shows a transition that might occur with small energy loss at a much larger radius to give m^{+*} and N^+. It would therefore appear that charge exchange is occurring in two distinct processes. Ions of the heavier rare gases do not, in general, show energy gain peaks; they are observed in a few cases only. With krypton ions, for example, using krypton as collision gas, a spectrum shown in Fig. 79 was obtained. The energy gain peak has a maximum height at a gain of 1.7 eV and a tail corresponding to twice this energy. The larger peak shows an energy loss of $\sim$3.2 eV. The transition $Kr^{2+} \rightarrow Kr^{+*}$ (lowest excited state) releases 11.0 eV, and the ionization potential of krypton is 14.0 eV, a difference of 3.0 eV.

It is found that when a diatomic collision gas such as N_2 is employed, a single intense peak very near to zero energy is observed. Presumably, the large number of low-lying vibrational states in $N_2^{+\cdot}$ enable this ion to be formed with a much better chance of very good energy matching.

A second type of charge-exchange process can be studied if the voltage across the electric sector is reversed. Negative ions are formed from high kinetic energy positive ions in the presence of collision gas by either one or other of the processes

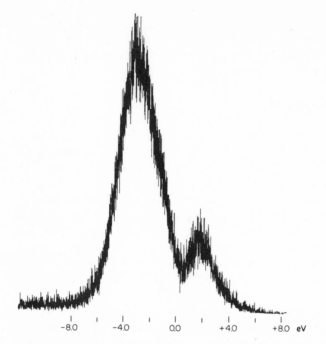

Fig. 79. Kinetic energy spectrum due to the charge-exchange reaction (20/11) for Kr^{2+} on Kr.

$$m^+ + N \rightarrow m^- + N^{2+}$$

or

$$m^+ + N \rightarrow m + N^+$$

followed by

$$m + N \rightarrow m^- + N^+$$

Reactions of these kinds have been studied in detail by Durup and others[6, 236] and fine structure has been observed in the kinetic energy loss spectra of the negatively charged product ions. This fine structure has been correlated with the energy necessary to form the product positive ions in various electronic states. Most of the work has, so far, been carried out using rare gases or oxygen, nitrogen or nitric oxide as collision gas and simple ions m^+ such as protons and $O_2^{+\cdot}$. Negatively charged product ions can also readily be observed when organic positive ions (especially those containing halogen atoms) are made to undergo charge exchange[160].

COLLISION-INDUCED FRAGMENTATION

The interaction of a polyatomic ion of high kinetic energy and a neutral atom or molecule may lead to fragmentation of the ion. If this process occurs in the field-free region of a mass spectrometer, it will give rise to a peak which is similar in appearance to a normal metastable peak. The formation of such peaks, like electron impact-induced fragmentation, involves two separate steps, excitation and then fragmentation. The excitation step involves energy transfer mainly to the electrons of the ion; there is very little momentum transfer[177, 232]. Indeed, a close parallel sometimes exists between reactions induced by collision and those which occur in the ion source, as has been pointed out by several authors[154, 232]. Partly on the basis of qualitative results of this type and in part on the basis of a more detailed analysis, including comparisons with theoretical predictions for small ions, it is believed that Franck—Condon type transitions occur in the excitation step[209].

Durup[102] has considered in detail several different mechanisms of conversion of kinetic to internal energy in an ion–molecule collision. The major process at energies above 1 keV is a vertical electronic excitation which occurs essentially without momentum transfer to the neutral species. A competitive process which becomes more important at low energies is an adiabatic vibrational excitation process. Simultaneous excitation of the target occurs in this reaction which involves considerable momentum transfer. It has been shown, in simple systems at least, that the nature of the collision gas can determine whether the electronic or the vibrational excitation process is favored. Small targets such as helium produce largely electronic excitation while large targets such as xenon tend to favor the vibrational excitation process. Provided the electronic excitation mechanism is operative, one can ignore the small amount of energy acquired by the collision gas and determine directly, by monitoring product ions which have been scattered through a small angle centered about zero, the amount of kinetic energy (Q) of the reactant ion which has been converted into internal energy. The experimental conditions used in the IKES work described in this book are appropriate to the electronic excitation mechanism and the position of the peak in the IKE spectrum gives Q' which is approximately equal to Q. Because ions scattered through appreciable angles are not accepted, Q' is always a small quantity relative to the ion energy and usually it is less than 10 eV.

The generally small kinetic energy losses in collisions leading to fragmentation may make the exact positions of peaks due to collision-induced processes difficult to distinguish from those due to metastable peaks. The amplification factor (p. 63) which is responsible for the breadths and the variety of shapes of metastable peaks, also operates for the collision-induced peaks. This factor so broadens the peak that any spread in the energy loss Q' will almost always make a far smaller contribution to the peak width than will energy release, T, (whether or not a range of T values be involved). Thus, peak position gives, quite simply, the energy loss, while peak width gives the energy release. The validity of this contention is shown by the fact that peaks solely due to a single (collision-induced) fragmentation are always observed to be symmetrical about their center.

It is important that peaks due to collision-induced fragmentations be distinguished from metastable peaks, a task made difficult by the similarities in shape and position already noted. It has not been the general practice among mass spectroscopists to attempt to distinguish between these processes in the spectra of organic ions, yet even the most qualitative use of metastable peaks as a guide to fragmentation pathways assumes the unimolecularity of the process under consideration. For these reasons, combined with the fact that collision-induced dissociations have relatively high cross-sections, it is believed that a sizeable fraction of the processes which have been reported as reactions of metastable ions are, in fact, collision-induced fragmentations.

There are two simple ways in which the distinction between unimolecular and bimolecular reactions can be made. First, the pressure in the field-free region can be varied by adding collision gas. A plot of signal strength *versus* pressure should be linear and pass through zero if the reaction is collision-induced. A lack of pressure dependence [except that scattering, charge exchange and other reactions usually decrease the signal at pressures above 10^{-2} to 10^{-3} Pa ($\sim 10^{-4}$ to 10^{-5} torr)] indicates a purely unimolecular reaction. A non-zero intercept combined with a linear pressure dependence indicates that both types of reaction are occurring. Second, if an instrument of moderate energy resolution is available, a good indication of whether or not a given signal is due to a metastable ion decomposition comes from the behavior of the peak width as a function of collision gas pressure. For example, methane shows a peak for the reaction $16^+ \rightarrow 15^+ + 1$ which has a width corresponding to the release of 5 meV kinetic energy at low analyzer pressures (10^{-5} Pa, $\sim 10^{-7}$ torr). As methane is admitted as collision gas, the signal increases greatly in abundance, but also increases in width, corresponding to release of ~ 110 meV. This result is to be contrasted with that found for the reaction $CH_3^+ \rightarrow CH_2^{+\cdot} + H^\cdot$ in methane[50]. In this case, the peak again increases in abundance but its width (corresponding to ~ 210 meV energy release) remains constant. These results indicate that $CH_4^{+\cdot}$ loses $H^\cdot$ by both a unimolecular and by a collision-induced process, while CH_3^+ undergoes only a collision-induced reaction.

Because the cross-sections for collision-induced fragmentation are high and because many more stable ions pass through the analyser of a mass spectrometer than do unstable ions, excitation of these ions to states from which they can fragment is common. Thus, collision-induced reactions are much more numerous than unimolecular transitions (see Chapter 5). In addition, transitions which occur unimolecularly are often enhanced by the addition of collision gas.

The occurrence of both a unimolecular and a collision-induced process corresponding to the same reaction leads to the observation of overlapping peaks. Unlike the composite peaks due to two competitive unimolecular reactions (see Fig. 61), the peak centers in this case do not exactly match. This is also shown in the case of loss of $H^\cdot$ from dimethyl ether in Fig. 80. The metastable (sharp) peak

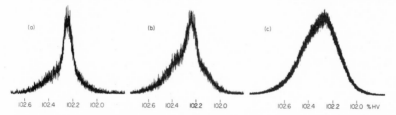

Fig. 80. Metastable peak plotted by the HV scan technique due to loss of $H^\cdot$ from the molecular ion of dimethyl ether. As the pressure in the field-free region is increased, the collision-induced component becomes predominant.

serves as a valuable reference point in determining Q' for the collision-induced process. If the peaks are well separated, Q' can be determined accurate to 0.5 eV; if there is no unimolecular transition to serve as a reference, the error will usually be larger.

The dimethyl ether molecular ion undergoes $H^\cdot$ loss in both a unimolecular and in a collision-induced reaction[57]. The energy release accompanying the unimolecular reaction is only 4 meV and the corresponding narrow metastable peak is shown in Fig. 80(a). Even at the pressure of 1.5×10^{-5} Pa (2×10^{-7} torr) used in this experiment, a second broad peak shifted to lower energy can be seen. As air is introduced as collision gas and the pressure in the field-free region is raised, this peak increases in intensity as shown in (b) and finally becomes dominant, as shown in Fig. 80(c). The behavior observed in dimethyl ether, *viz.* that the energy release accompanying the collision-induced reaction is greater than that accompanying the unimolecular process, is common. Indeed, the limited number of measurements done so far on the widths of collision-induced peaks show that they are almost invariably broader than their unimolecular counterparts, in agreement with the idea that fragmentation occurs from the same electronic state having different amounts of vibrational/rotational energy. However, it is expected that isolated electronic states should also be accessible on collision and that fragmentation from such high-energy states might result in less kinetic energy release than for unimolecular reaction (a smaller value of T, for example, could be the result of a smaller reverse activation energy due to formation of an electronically excited form of one of the products). This is apparently the case for the reaction

$$CH_2 = \overset{+}{O}H \rightarrow CHO^+ + H_2$$

in methanol, which is illustrated in Fig. 81. Here the average kinetic energy release is *ca.* 50 meV for the collision-induced reaction and 174 meV in the unimolecular process. The most probable loss of ion kinetic energy upon collision, as measured from the displacement of the resulting peak upon fragmentation, is 12 eV. This methanol reaction was also examined in the second field-free region. In the presence of collision gas the metastable peak occurring in the mass

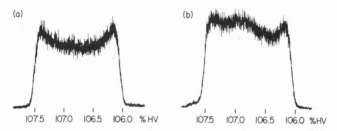

(a)

(b)

| 107.5 | 107.0 | 106.5 | 106.0 % HV |

| 107.5 | 107.0 | 106.5 | 106.0 %HV |

Fig. 81. Unimolecular and collision components of the peak due to loss of H_2 from m/e 31 in methanol. The collision-induced peak is visible in part (b) as a narrower signal which is centered at higher HV. At pressures higher than those used for (b) the narrower peak becomes dominant.

spectrum was shifted to lower mass by 0.026u from the position (27.118u) calculated for the reaction $31.018^+ \rightarrow 29.003^+ + 2.016$. This shift corresponds to a kinetic energy loss by the reactant ion of 11 eV, in good agreement with the value measured in the first field-free region. The selection of internal energies of metastable ions effected by the instrument means that the excess energy ϵ^+ is low in the normal experimental arrangement. However, no such selection applies to collision-induced fragmentation which might yield ions fragmenting immediately upon collision or as much as several microseconds later. The possibility of sampling ions with high excess energies accounts for the typically larger kinetic energy releases recorded for these processes. It also accounts for another generalization which can be made regarding peaks due to collision-induced reactions. These peaks, including those for the dimethyl ether and methanol reactions just illustrated, typically have sides of much shallower slope than those due to the corresponding reactions of metastable ions. Hence, a wider range of kinetic energies is released in the collision-induced reactions.

It appears that when organic ions of energy $\sim$ 4 keV impinge upon argon as collision gas, the amount of internal energy acquired by the ion is less than the average amount typically transferred by electron impact[199]. In other words, by examining collision-induced reactions, one can study ions which have internal energies which are intermediate between the higher-energy ions which react in the source and the low-energy metastable ions. This hypothesis has been used by McLafferty in several ways, for example, in studying the effect of internal energy upon isotope effects[147] and the results obtained argue for its validity. The use of collision-induced reactions to follow isotope scrambling, however,[194] is fraught with difficulties because the observed effects are due to the sum of the reactions which occur in the long-lived low-energy ion prior to excitation and those which occur in the short-lived higher-energy species in the interval between excitation and fragmentation.

Collision-induced fragmentation of small ions (that is, diatomics and triatomics) has been studied extensively. The reaction $H_2^{+\cdot} \rightarrow H^+ + H^{\cdot}$ alone has been the subject of numerous studies, many listed in Durup's paper[101]. The resulting peak shape is shown in Fig. 82. The narrow spike corresponds to the vibrational excitation process discussed earlier in this section and the broad peak to the electronic excitation. Vibrational fine structure has occasionally been seen in peaks due to fragmentation after collisional excitation. The reaction $H_3^+ \rightarrow H^+ + H_2$ provides a case in point, different levels among the

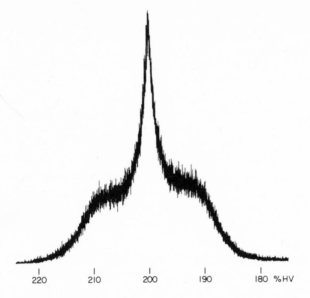

Fig. 82. Collision-induced decomposition $H_2^{+\cdot} \rightarrow H^+ + H^\cdot$ plotted by the HV scan technique. The narrow component arises by a vibrational excitation mechanism and the broad component by electronic excitation of $H_2^{+\cdot}$.

upper vibrational states of H_2 being populated and thus causing discrete steps in the kinetic energy released in the reaction[87]

Chapter 5

Analytical applications

"But wot's that you're a doin' of? Pursuit of knowledge under difficulties, Sammy? "

Dickens, *Pickwick Papers*

Metastable ions have been used in many different ways to obtain information about organic and inorganic compounds. One aspect of these studies is concerned with ion structures and properties and another with analytical data. Although it is difficult in some cases to separate these two areas, work concerned mainly with analytical applications of metastable ions will be discussed in this chapter. All studies made for the purpose of gaining information about the neutral compound are, therefore, included. However, studies designed mainly to provide information concerning the ionized compound, such as ion structures, thermodynamics, etc. are discussed elsewhere.

Many analytical applications of metastable ions are concerned with the elucidation of the fragmentation pathways of compounds in order to gain information on molecular structure. Although this is one of the primary uses of metastable ions in analysis, this chapter will emphasize other types of information obtainable through the study of metastable ions. These include the analysis of mixtures of compounds, isotopic distributions and the identification of isomers. Some topics, such as rearrangement processes, will be mentioned only insofar as their application to the determination of molecular structure is concerned.

ELUCIDATION OF FRAGMENTATION PATHWAYS

One of the early applications of mass spectrometry was its use as an analytical tool through which molecular structure could be determined. To accomplish this, fragmentation patterns were rationalized by postulating reasonable ion structures and from this information, molecular structure was inferred, sometimes with the aid of isotopic labeling. The discovery of metastable ions, however, provided another approach through which pairs of ions in a mass spectrum could be linked to elucidate partial or complete fragmentation pathways. From a structural point of view, this is important because it allows the chemist to determine the successive losses of fragments starting with the molecular ion. The observation of a metastable transition does not provide unequivocal proof of a one-step reaction, since, if a fast reaction follows a slow one, a peak will be seen for combined loss of the two fragments as well as one for loss of the first fragment alone. In all cases so far observed however, a separate peak for the second reaction is also seen.

The partial fragmentation pattern of bis-pentafluorophenylamine has been worked out by recording metastable peaks for each decomposition[168], as shown in Fig. 83. All ions in the figure are interconnected by reactions of metastable ions; only by following these reac-

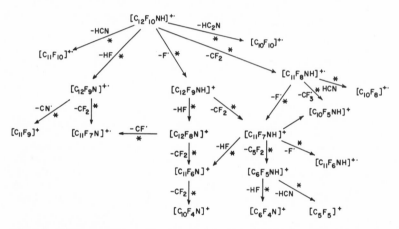

Fig. 83. Fragmentation pattern of bis-pentafluorophenylamine elucidated by recording metastable peaks in the mass spectrum.

tion sequences can structural information be deduced. Even high-resolution mass spectrometry, which can provide the elemental composition of every ion, gives no information about the reaction sequence by which the ion is formed. Thus, the appearance of a $C_6F_4N^+$ ion in Fig. 83 is of little structural significance without the information about the four successive steps by which it can be formed.

The detailed fragmentation pattern of nonan-4-one has also been elucidated through the study of metastable ions in a double-focusing mass spectrometer[105]. In this work, metastable ions decomposing in both the first and second field-free regions were recorded by using IKES and HV scans. The study of reactions in the first field-free region leads to the unambiguous assignment of metastable transitions. If the diffuse metastable peaks occurring in the mass spectrum alone are used, multiple assignments would occur in some cases.

The major features of the fragmentation pattern are shown in Fig. 84. The first important factor to be noted is that more fragmentations involving the molecular ion as parent ion can be seen than for any other ion. The most important contribution of mass spectro-

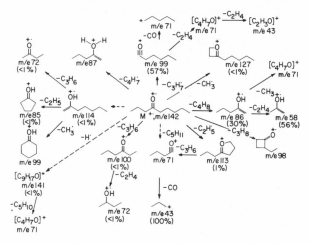

Fig. 84. Fragmentation pattern of nonan-4-one elucidated by ion kinetic energy spectrometry. The broken arrows indicate transitions for which no metastable ion decompositions were recorded.

metry to structure elucidation in organic chemistry has been the assignment of the molecular weight. And yet this is often done on the tenuous evidence of a small mass peak and the indirect evidence that the masses that appear to be lost from this ion are reasonable. On the other hand, metastable peaks generally point much more strongly to the molecular ion; both the number of transitions and the direct evidence of the masses lost in single-step reactions make the assignment of this important ion much more certain.

The large number of metastable transitions that can, generally, be observed (often larger than the number of normal mass peaks), is in no way a disadvantage. It can be seen from Fig. 84 that the most abundant metastable peaks are for reactions such as α-cleavage of the short chain and McLafferty rearrangement in the long chain both of which are very important reactions for structural diagnosis. The less abundant peaks are also important; they indicate the mechanism by which low probability single-step reactions can occur and also point out the multi-step fragmentations that lead to unexpected ions. When the decomposition pathways are clarified in this way, peaks of low abundance can become useful in structural assignments.

For IKE or HV scans, only a voltage is varied in order to scan the spectra. Programmable power supplies can be used with a small computer to switch rapidly between parts of the scan where peaks are expected. Very weak peaks can be enhanced by highly reproducible repetitive scanning; the linking together of pairs of mass numbers corresponding to the parent and daughter ions of single-step transitions enables individual fragmentation sequences to be characterized and opens the way to the automated interpretation of mass spectra. In contrast, mass spectra are commonly produced using magnetic sector instruments in which case scans are not reproducible and the construction of an accurate mass scale necessitates a reference compound as internal standard. Many of these difficulties can be circumvented by using a quadrupole in which frequency determines the mass scale accurately, but even here, the advantage of seeing individual fragmentation steps is lost.

In large molecules, fragmentation patterns become increasingly complex and the elucidation of molecular structure from them becomes more difficult. It is possible to obtain structural information more readily by coupling metastable ion data with high-resolution mass spectrometry. In this approach[16], the magnet current is adjusted manually to collect one of the daughter ions. An HV scan then determines the masses of its precursors. This procedure is repeated at magnet current settings corresponding to each of the daughter ion species in turn. The data read into a computer consists of the nominal mass of each fragment ion, the accelerating voltage to collect each parent, the accurate mass of each ion studied and the maximum number of heteroatoms lost in the neutral species in each step. All possible fragmentation pathways are identified from the low-resolution data. Some transitions will occur both by single- and multiple-step pathways; the latter contain more structural information and the shorter pathway is considered redundant for the purposes of the computer program and is eliminated. For the remaining transitions, the accurate masses of the daughter ions formed in the ion source are assumed to represent the formulae of those daughters formed in the field-free region and are subtracted from that of the molecular ion, thereby establishing the elemental composition of the neutral species lost. For example, a peptide derivative of molecular weight 638 was studied in this manner. A redundancy check of the metastable ion data reduced 81 different observed fragmentation pathways to a useful 33. Several pathways began with the transition $638 \rightarrow 494$, but were duplicated by others that began with $638 \rightarrow 623 \rightarrow 494$. Thus, only the latter pathway was retained. By such processes used in conjunction with the high-resolution data one fragmentation pathway was ascertained to be

$$C_{35}H_{66}N_4O_6{}^{+\cdot} \xrightarrow{-CH_3{}^{\cdot}} C_{34}H_{63}N_4O_6{}^{+} \xrightarrow{-C_6H_{11}NO_2} C_{28}H_{52}N_3O_4{}^{+}$$

$$\downarrow -C_2H_3NO$$

$$C_{20}H_{40}NO^{+} \xleftarrow{-CO} C_{21}H_{40}NO_2{}^{+} \xleftarrow{-C_5H_9NO} C_{26}H_{49}N_2O_3{}^{+}$$

$$\downarrow -C_2H_5N$$

$$C_{18}H_{35}O^{+}$$

It is reasonable to assume that the original molecule is built up of some arrangement of these neutral fragments, remembering that skeletal rearrangement can occur in the individual steps. In the above example, the six fragments shown could give rise to 180 different possible structures. However, since any arrangment must satisfy other similarly deduced pathways, most of the possible structures can be ruled out. In this way, consideration of only six fragmentation sequences resulted in only two possible arrangements of all the fragments into alternative molecular structures. This approach attempts to solve one of the major problems of data analysis in mass spectrometry by pooling different types of data. It is anticipated that with the increased sensitivity now available for studying metastable ions and the consequent increase in the number of transitions that can be detected, the elucidation of molecular structures in this way will be facilitated.

This section concludes with a brief discussion of three techniques which have potential in supplementing the structural information which can be deduced from metastable singly charged ions formed by electron impact. The techniques are relatively new and their analytical applications are embryonic.

It has been shown[264] that doubly charged ion mass spectra can sometimes be used to distinguish isomers when this cannot be done using normal mass spectra. With the development of the charge-exchange method[10, 49, 55] of recording doubly charged ion mass spectra free of interferences from singly charged ions, the analytical potential of such spectra has been much increased. Spectra can be obtained at high sensitivity and a study on amines[11] indicated the existence of strong spectrum/structure correlations. It is not yet clear to what extent this technique will add to the structural information available from a normal mass spectrum.

Collision-induced dissociations which occur in a field-free region of the mass spectrometer give rise to peaks which resemble those due to metastable ion fragmentations. Such processes have two important advantages in an analytical sense. First, they often give rise to more abundant peaks than those resulting from unimolecular reactions. Second, many more reactions occur on collision than occur unimolecularly. In approximate terms, collisional excitation produces species in a variety of electronic states and fragmentations from these states are sampled from zero time. The increase in the number of reactions which can be identified assists in deriving a molecular structure consonant with the observed behavior. These points may be illustrated by the partial mass spectrum of pyrazine shown in Fig. 85.

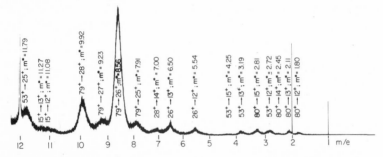

Fig. 85. Partial mass spectrum of pyrazine taken in the presence of collision gas in the second field-free region of a double-focusing mass spectrometer.

The spectrum is taken in the presence of collision gas in the second field-free region and the part of the spectrum shown, from m/e 1 to m/e 12, is normally the least valuable part of any mass spectrum. The large number of transitions observed and the frequent occurrence of 80^+, 79^+ and 53^+ as parent ions speaks for the analytical potential of this procedure. (Pyrazine has molecular weight 80 and undergoes primary fragmentation reactions to give abundant m/e 79 and m/e 53 ions.)

Metastable ions formed by field ionization can provide information which is valuable in interpreting fragmentation patterns and, hence, in deducing molecular structure. As is the case for electron impact, abundant metastable peaks are associated with reactions of low activation energy, often rearrangement reactions. The fact that rearrangement reactions do not usually give rise to normal daughter ions in field ionization (FI) mass spectra allows these processes to be identified by comparison of metastable and daughter ion peaks. This principle has been helpful in correlating FI spectra and molecular structure[2 3].

ANALYSIS OF MIXTURES

Metastable ions can be invaluable in the analysis of mixtures of compounds because they link pairs of ions in the spectrum and thus prove that both these ions come from a single component. As an example of the use of metastable ions in such a situation, consider the chemical reaction[4 1]

In addition to the major product shown above, there was also an impurity assumed to be a by-product of the reaction. The molecular formula of the impurity was established by accurate mass measurement to be $C_{19}H_{15}N_5O$. A series of metastable peaks then showed the fragmentation pathway of this ion to be

$$C_{19}H_{15}N_5O^{+\cdot} \xrightarrow{-C_6H_5N_2^{\cdot}} C_{13}H_{10}N_3O^+ \xrightarrow{-C_2H_3N} C_{11}H_7N_2O^+$$

$$\downarrow -CO$$

$$C_8H_5^+ \xleftarrow{-HCN} C_9H_6N^+ \xleftarrow{-HCN} C_{10}H_7N_2^+$$

Assuming the $C_8H_5^+$ ion to be the remains of the naphthalene nucleus, the loss of two successive molecules of HCN indicates that the two nitrogen atoms were still attached to the ring system in the impurity, while the loss of $C_6H_5N_2^{\cdot}$ from the molecular ion shows the phenylazo group to be present. The single oxygen atom present is lost as CO. The remaining fragment C_2H_3N appears to be lost as an entity, and NMR shows the impurity to contain a methyl group. In keeping with the structure of the main reaction product, the structure of the impurity was suggested to be

Thus, even though this compound was present as the minor component of a mixture, metastable ions were of fundamental importance in enabling its structure to be determined.

Another example of the use of metastables in the analysis of mixtures of unknowns is provided by the problem of peptide structure determination. Partial peptide sequence information can be obtained on mixtures using a combination of exact mass, partial vaporization and metastable ion data[195]. A high-resolution mass spectrum was taken and a computer program used to identify the possible sequences consistent with these exact mass measurements. A total of eleven possible sequences was found involving five different molecular compositions.

A.   Ac-gly-OMe
B.   Ac-gly-gly-val-OMe
C.   Ac-gly-ala-gly-OMe
D.   Ac-giy-ala-leu-OMe
E.   Ac-gly-val-gly-OMe
F.   Ac-ala-gly-gly-OMe
G.   Ac-ala-gly-leu-OMe
H.   Ac-ala-ala-ala-OMe
I.   Ac-ala-ala-val-OMe
J.   Ac-val-gly-gly-OMe
K.   Ac-met-phe-gly-OMe

Nine metastables from the mixture were consistent only with the presence of K, eight with J, and fourteen with D. No metastables were detected which would be unique for other postulated sequences. Thus, the presence of D, J and K was correctly ascertained. This represents a marked improvement over the study of stable ions in the mass spectrum alone, by which the above identification could not have been made.

Metastable ions have also been used to identify mixtures of thiohydantoin derivatives of amino acids[253]. These derivatives are obtained as the product of the Edman degradation in the sequencing of amino acids in proteins. The mass spectrum of each of the thiohydantoin amino acids contains at least one metastable transition which is unique to that compound. For example, glycine phenylthiohydantoin

was found to undergo two decompositions of metastable ions which are unique, one giving a peak at m/e 140.08 corresponding to loss of CO from the molecular ion, and the second at m/e 114.44, corresponding to loss of HCN from the $(M—CO)^{+\cdot}$ ion. Thus, these unique metastable peaks can be used to identify one or more of these derivatives in mixtures. This is particularly important in the identification of glycine in mixtures, since the phenylthiohydantoin glycine molecular ion is also a major fragment ion of many of the other components in the mixture.

One of the basic ways in which a mass spectrometer is used is for determining the molecular weight of a sample. The value of a fragmentation map in indicating the molecular ion has already been stressed earlier in the chapter. In a mixture, however, it is often difficult to pick out any molecular ions except those due to the component of highest molecular weight. Moreover, this ion may not arise from the major component which may be missed due to the presence of a homolog of molecular weight, say, 28 mass units higher. One way of gaining further information concerning molecular ions is to make use of the fact that many molecular ions of organic compounds show a peak due to loss of a single hydrogen atom. This peak could be coupled to the molecular ion by the metastable peak appearing at a mass-to-charge ratio of approximately $(M—2)$. Unfortunately, this metastable peak is often narrow and lost in the skirts of the fragment peak $(M—2)^{+\cdot}$. In such cases, the continuum from $M^{+\cdot}$ to m* (discussed in Chapter 3) can generally be observed. The presence of this continuum is strong evidence that the higher mass from which it extends corresponds to a molecular ion. Even in the spectrum of a complex mixture, the continuum is diagnostic of the presence of a molecular ion. Figure 86 shows part of a mass spectrum in a region overlaid by fragment peaks and half masses from higher molecular weight compounds. A continuum can clearly be seen to extend from the peak at mass-to-charge ratio 78 down to that at 76, indicating that 78 is probably the molecular weight of a constituent of the

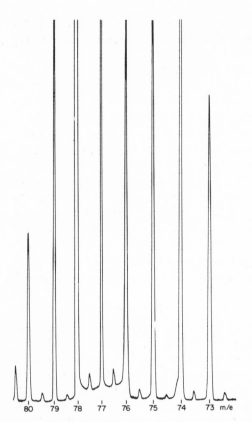

Fig. 86. Continuum between m/e 78 and m/e 76 in the mass spectrum of a mixture containing benzene, due to the transition $78^+ \rightarrow 77^+$ occurring in the magnetic sector.

mixture and that the continuum is due to the transition $78^+ \rightarrow 77^+$. By looking along the baseline, a further weak continuum can be seen to extend upwards from mass 74 at which position a metastable peak can be observed. This continuum is, thus, due to the transition $78^+ \rightarrow 76^+$, providing further evidence that 78 corresponds to a molecular ion.

Other applications of the use of metastable ions in the analysis of mixtures of isotopes and of structural and geometric isomers will be discussed in later sections of this chapter.

REARRANGEMENT PROCESSES

The study of metastable ions has aided in the interpretation of mass spectra through the elucidation of rearrangement processes. An understanding of the conditions and structural factors which allow these processes is vital to the extrapolation of mass spectral data back to the un-ionized form of the molecule. It is becoming clear that the electron-impact-induced fragmentation of organic molecules may be accompanied by a great many rearrangement processes. In many cases, this involves migrations and exchanges of hydrogen atoms between portions of an ion. In other cases, groups such as methyl, phenyl, benzyl, etc. can migrate. Metastable ions are of value here because they can be used to specify the ion from which the rearrangement occurred.

The study of hydrogen rearrangment processes in many systems, particularly in aromatic molecular ions, has been expedited by observations of metastable peaks. Studies based on stable fragment ions are often complicated by the occurrence of ions of different composition at a single mass and also by the fact that the precursor of a fragment ion of given elemental composition is not defined. As an illustration, consider the decomposition of thiophen and its deuteriated analogs[274]. The most abundant fragment ion in the 70 eV mass spectrum of thiophen is due to an $(M—C_2H_2)^{+\cdot}$ species and an abundant metastable peak is found for this transition (m/e 84 $\rightarrow$ 58) at m/e 40.0. 2-d_1-Thiophen undergoes the transitions for loss of C_2H_2 and C_2HD with equal probability, as established by two metastable peaks of equal intensity at m/e 40.9 and 39.6. Both 2,5-d_2-thiophen and 2,3-d_2-thiophen undergo transitions for loss of C_2H_2, C_2HD and C_2D_2 to give metastable peaks at m/e 41.8, 40.5, and 39.1 in the intensity ratio 1:4:1. These results demonstrate hydrogen scrambling in the form of the molecular ion which loses acetylene. Direct measurement of stable ions would not have provided unambiguous data in such an analysis.

It must be emphasized that the effect of the change in time scale involved in observing metastable ions rather than stable fragment ions is to sample ions of lower internal energy. In principle, the reactions undergone by these ions might differ substantially from those undergone by higher-energy ions. In practice, the chief difference lies in the degree of randomization which precedes fragmentation. An increase in the degree of hydrogen rearrangement with ion lifetime is a general phenomenon and has been illustrated by recent work with various compounds[146,280]. A detailed discussion of these studies, however, falls outside the scope of this chapter.

Metastable ions have been used to describe the mechanism of the rearrangement process in which H_2O and $COOH^\cdot$ are lost from the molecular ion of benzyl benzoate[61]. This rearrangement process is particularly diagnostic of chemical structure, since its isomers, phenyl phenylacetate and α-phenoxyacetophenone, do not show these losses.

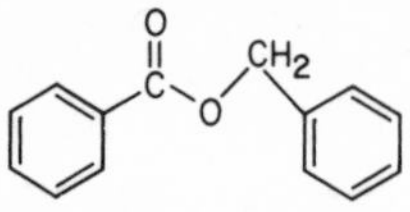

benzyl benzoate

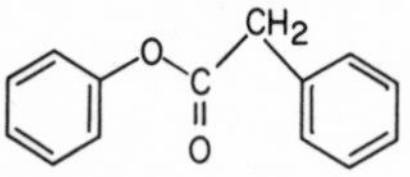

phenyl phenylacetate

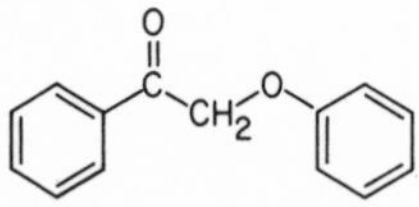

α-phenoxyacetophenone

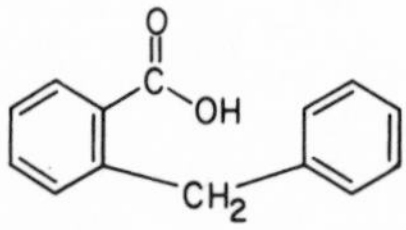

α-phenyl-o-toluic acid

However, one isomer, α-phenyl-o-toluic acid, also shows loss of H_2O and $COOH^{\cdot}$, although its structure differs considerably from that of benzyl benzoate. In order to detail the mechanism of this rearrangement, isotopically labeled benzyl benzoates were prepared. From these, the probabilities of loss of H_2O, HDO, D_2O or $H_2{}^{18}O$ were calculated for the appropriately labeled compounds. The rearrangement and subsequent fragmentation of the molecular ion of benzyl benzoate was proposed to be

The first step involves the abstraction of one of the ortho hydrogens by the carbonyl group, followed by a rearrangement to form an ion identical to the molecular ion of α-phenyl-o-toluic acid. This ion can then lose H_2O followed by loss of CO, as verified by metastable transitions, or it can lose the carbohydroxy radical. Results from the variously labeled compounds showed that the two benzyl ortho hydrogens, the methylene hydrogens and one of the benzoate ortho hydrogens are involved in loss of water. If we assume that these five hydrogens equilibrate, then we would expect the ratio of loss of hydrogens from the ortho positions on the benzoate moiety: methylene hydrogens : hydrogens from the ortho positions of the benzyl group to be 20:40:40. The experimentally measured ratios derived from metastable ion abundances were 20:36:44, in good agreement. It can thus be seen that the metastable transitions involving loss of H_2O and COOH˙ are of considerable analytical importance in trying to distinguish isomers. Further elucidation of the mechanisms of rearrangement reactions should allow extension of this general analytical approach.

ISOTOPE MEASUREMENTS

One of the most useful applications of metastable ions in stable isotope measurement lies in the analysis of isotopic mixtures. Since metastable ions link parent and daughter ions in unimolecular transitions, decompositions which occur from different parent ions yet give the same daughter ion can be distinguished. For example, in a recent study, $1,2\text{-}^{13}C_2\text{-}3,4,5,6\text{-}d_4$-benzene was prepared in order to determine the extent of carbon and hydrogen randomization following electron-impact ionization[60]. The sample was an isotopic mixture of 30% $^{13}C_2D_4$, 46% $^{13}C_1D_4$, 18% $^{13}C_0D_4$, 3% $^{13}C_2D_3$, 4% $^{13}C_1D_3$ and 1% $^{13}C_0D_3$. The molecular weights of these components are 84, 83, 82, 83, 82 and 81, respectively. Fragment ions of a particular mass may be derived from more than one of these components. A high-resolution mass spectrum can provide data on the elemental compositions of the fragment ion, but gives no information on the precursors from which they arise. Metastable peaks, however, link parent and daughter ions, at least to the extent that parent ions of different integral masses can be distinguished.

For example, in the present case, for the m/e 55 daughter ion, peaks from molecular species m/e 84, 83, 82 and 81 appear in an HV scan at 84/55, 83/55, 82/55 and 81/55 times the accelerating voltage, respectively (Fig. 87). Of course, different isotopic species

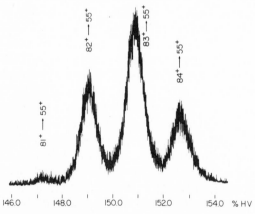

Fig. 87. Separate metastable ion fragmentations observed for molecular ions of labeled benzene having different isotopic compositions. The product ions were of mass-to-charge ratio 55 in each case and the peaks were recorded by the HV scan technique.

having the same nominal mass, such as $^{13}C_2D_3$ and $^{13}C_1D_4$, still would not be distinguishable. On the other hand, there may be cases where a single isotopically labeled parent ion could decompose to give more than one daughter ion. In the benzene study, the only molecular species of m/e 84 was the fully labeled $^{13}C_2D_4$ species and this decomposes into daughter species isotopically labeled in different ways. This situation too, can be analyzed using metastable ion decompositions, particularly if the MIKES technique is used. These examples once again exemplify the complementary nature of the information obtained from high-resolution and metastable ion studies.

Metastable ions have also been used for the determination of isotopic incorporation, for example in deuteriated toluene[52]. The difficulty encountered when peaks in the mass spectrum of this and many other organic molecules are analyzed is a result of the presence of both an $M^{+\cdot}$ and $(M-1)^{+}$ ion. Analyses of partially deuteriated compounds are then rendered difficult and inaccurate because of the large and often uncertain corrections which have to be made for the presence of the fragment ions. Thus, for toluene, a peak occurs at m/e 91 in the mass spectrum due to the transition

$$C_7H_8^{+\cdot} \rightarrow C_7H_7^{+} + H^{\cdot}$$

In the case of the analysis of a mono-deuteriated species, the ion at m/e 92 can either be due to the fragment ion as shown above, or an undeuteriated molecular ion. Now consider metastable ions. A metastable peak appears in the mass spectrum of unlabeled toluene at m/e 90 for the above transition. To transmit the fragment ion of mass 91 (which gives rise to the metastable peak at mass 90), the electric sector is tuned to a voltage equal to a fraction 91/92 of its normal value and the magnet current is scanned. If this is done successively for values of 90/91, 89/90, etc. other metastable transitions can be plotted in which ions of mass 92, 91, 90, etc. lose a single hydrogen atom. Such a scan is shown in Fig. 88 and it can be seen that there are no peaks (except the isotope peak at mass 91) greater in intensity than $\sim$ 2% of the peak due to $92^{+} \rightarrow 91^{+}$. If a monodeuteriotoluene is examined in the same way, the peak at mass 91 (corresponding to $93^{+} \rightarrow 92^{+}$) is again by far the largest peak. The peaks at masses 90 and 91, after correcting for naturally occurring isotopes, are proportional in the spectrum of the mixture to the molar amounts of toluene and monodeuteriotoluene. The proportionality constants, which are close to unity, can be determined simply by adding a known amount of toluene to the mixture and remeasuring the peak areas of the metastable peaks at masses 90 and

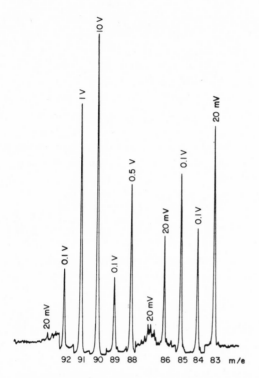

Fig. 88. Determination of deuterium incorporation in toluene by the IKES technique.

91. This method is quick and accurate and should be of wide application in analyses of deuteriated compounds.

IDENTIFICATION OF ISOMERS

The study of metastable ions, especially in the form of IKE spectra, can be of great value in the analysis of isomeric compounds whose mass spectra are nearly identical. The energy spectrum of each compound represents a specific pattern which is different for various compounds. For example, consider the IKE spectra of o-, m- and p-phenylenediamine[79] from electric sector voltage zero to E given in Fig. 89. Although the spectra are similar, some differences can be observed. The intensities and ratio of the peaks in the doublet at approximately $0.62E$ are significantly different in the three compounds and this is also true of the doublet at $0.51E$. In the ortho-isomer, the peak at $0.40E$ is considerably larger than those in the spectra of the other isomers, while in the para-isomer, the peak at

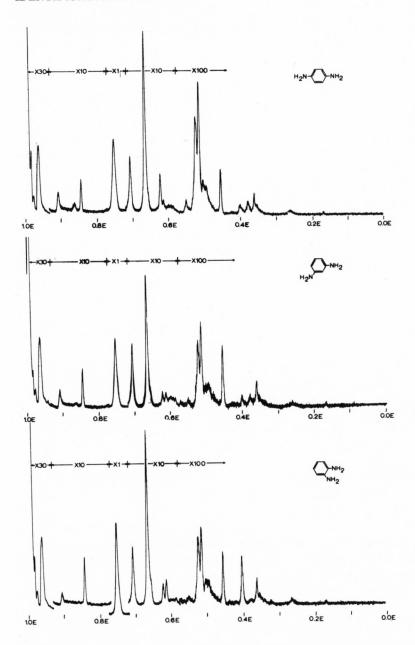

Fig. 89. IKE spectra of *o*-, *m*- and *p*-phenylenediamine illustrating their differences and similarities.

$0.86E$ is larger. Thus, these isomers can be distinguished on the basis of their IKE spectra. This is in contrast to their mass spectra which are identical.

Camphor and its isomer, *trans*-8-methylhydrindan-2-one, give almost identical mass spectra[80]. Their IKE spectra, on the other hand, show considerable differences. One of these differences is the presence of a peak at $0.72E$ for camphor corresponding to loss of ketene from the molecular ion and the absence of this peak in the isomeric ketone. Another example, this one of current analytical importance, is provided by the identification of isomeric polychloro-biphenyls using their IKE spectra[238]. This could not have been accomplished by using mass spectra alone.

Three isomers of $C_4H_4N_2$ have also been examined in regard to differences seen in metastable transitions which are unique for a particular isomer[59]. Thus, pyrazine, pyrimidine and pyridazine have been studied using IKES. In the spectra of these isomers, given in Fig. 90, it can be seen that pyridazine shows major differences compared with pyrazine and pyrimidine. The major fragment ion in pyridazine is due to loss of N_2 from the molecular ion, while for the other two isomers, HCN is lost. Thus, even though differences in the intensities of peaks at m/e 52 and 53 can be seen in the mass spectrum of pyridazine with respect to the others, the remainder of the spectra are very similar, unlike the IKE spectra which show major differences. Differences can also .be seen in the IKE spectra of pyrazine and pyrimidine. For example, the ratios of the two largest peaks in each spectrum are different; for pyrazine, the ratio of the $0.67E : 0.49E$ peaks is approximately 2:1, while the ratio of the same peak in pyrimidine is approximately 1:1. Also, the peak at approximately $0.66E$ has a second component for pyrimidine (seen in Fig. 90 as a slight shoulder halfway up the peak), but not for pyrazine. This is more easily seen in the appropriate portions of the IKE spectra of d_4-pyrazine and d_4-pyrimidine, shown in Fig. 91, in which the peaks for d_4-pyrimidine are nearly resolved as a result of the greater mass difference between the deuteriated fragments. For unlabeled pyrimidine, these peaks result from the transitions

$$C_4H_4N_2^{+\cdot} \rightarrow C_3H_3N^{+\cdot} + HCN$$

$$C_4H_3N_2^{+} \rightarrow C_3H_2N^{+} + HCN$$

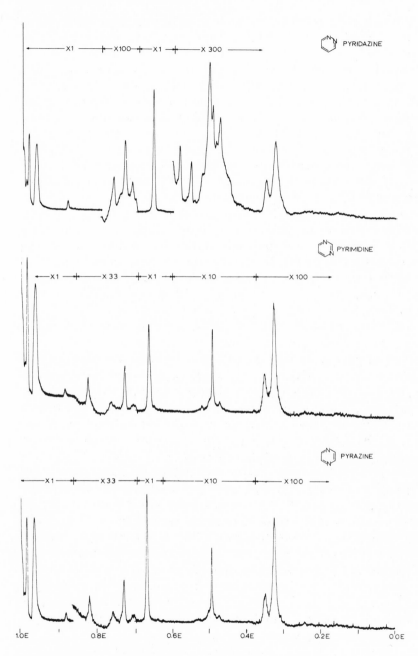

Fig. 90. IKE spectra of pyridazine, pyrimidine and pyrazine.

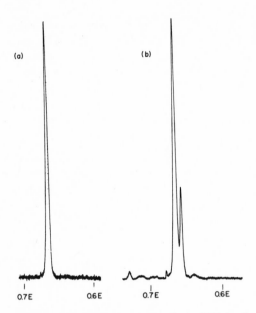

Fig. 91. Peak at approximately $0.67E$ in the IKE spectrum of (a) d_4-pyrazine and (b) d_4-pyrimidine.

The second does not occur for pyrazine. This serves to illustrate the sensitivity of IKE spectra towards differences in the structures of isomers. On the other hand, the largest difference in the mass spectra of these two isomers is the presence of the $C_4H_3N_2^+$ ion in pyrimidine with an intensity of 4% of that of the base peak, while this ion is less abundant in pyrazine. This result is consistent with, but less dramatic than, the differences in the IKE spectra.

Geometrical isomers have also been distinguished through the use of IKE spectra. Thus, *cis*- and *trans*-4-*tert*-butyl-cyclohexanol have different IKE spectra[80]. The *trans*-isomer shows abundant ions at approximately $0.58E$, $0.68E$ and $0.75E$, while these are either very small or absent in the *cis*-isomer. The determination of stereochemical structure by mass spectrometry involves a comparison of the relative extents to which diastereomers undergo a given elimination reaction[206]. This comparison has usually been made using daughter ion abundances, but since metastable transitions occur in lower-energy ions, metastable peaks will provide a more sensitive method of comparing their fragmentations.

A variation on the methods used for the identification of isomers from the properties of metastable ions is provided by considering the shapes of metastable peaks. This is illustrated by the identification of *m*- and *p*-nitroaniline using the metastable peaks for NO˙ loss from the molecular ion. The para-isomer gives a broad dish-topped peak and the meta-isomer yields a much narrower gaussian-shaped peak. The difference is related to the enthalpies of the product ions, resonance stabilization via a quininoid-type structure being possible only in the case of the *para*-isomer, as shown (compare p. 194).

Decompositions of metastable ions have been used to identify particular isomers present in mixtures of isomeric hydrocarbons. The abundances of the major metastable ions in the five isomeric hexanes, relative to the abundances of the fragment $C_6H_{14}^{+\cdot}$ ions in the mass spectra, were recorded using the HV scan technique[189]. At least one transition could be found for each isomer (except *n*-hexane) in which the metastable peak was unique to that isomer. For example, 2,3-dimethylbutane gave an abundant peak for the transition $86^+ \rightarrow 71^+$ and 2,2-dimethylbutane for $71^+ \rightarrow 43^+$.

Field-ionization techniques also have been employed to record primary metastable ion transitions in order to distinguish between isomeric mixtures. Three octadecane, six hexadecane and eighteen octane isomers have been examined in this way[269]. The field-ionization spectra of the octanes indicate that metastable ions can be used to analyze a mixture of these isomers by matrix analysis. As a result of the limited amount of fragmentation, the largest peaks above m/e 43 in these compounds, as well as in the paraffins, are frequently metastable peaks. For example, the largest such fragmentation peak in *n*-hexadecane is at m/e 126.4, corresponding to loss of 57 mass units from the molecular ion. The spectrum is thus relatively free of normal fragment ions, but metastable ions are of relatively high abundance. The relative abundances vary with the structures of the isomers and correlate with cracking patterns based on simple cleavage rules. This work also demonstrated the possibility of analysis of paraffin/olefin mixtures based on conversion of the olefins to alkylthiols followed by metastable ion studies with field ionization.

There are cases in which no differences in the IKE spectra of isomers can be detected. No differences have been seen for the *meta* and *para* isomers of fluorophenylacetylene or for 1-butene and *cis*-2-butene[80]. However, it has been suggested that the lack of difference in the case of the butenes should be attributed to a common structure for the molecular ion prior to decomposition in a metastable transition. In general, IKE spectra between *E* and *2E* (due to fragmentation of doubly charged ions to give singly charged ions) do not show any differences for different isomers. This is probably the result of the isomerization of the doubly charged molecular ions to a common structure.

EXACT MASS MEASUREMENTS

Accurate mass measurements of metastable peaks occuring in a mass spectrum have allowed the differentiation of two metastable transitions involving the same nominal masses. The accuracy which can be achieved is limited by the noise and peak shape. Measurements performed on fairly intense peaks show that results can be accurate to 250 ppm, which in some cases is sufficient to identify the transition[43]. For example, in the mass spectrum of aniline, metastable peaks appear at m/e values of 46.9 and 45.9. Two metastable transitions could possibly give rise to each of these peaks. However, on accurate mass measurement, a single metastable transition could be assigned in each case as shown in Table 4. Such measurements could be important in arriving at correct structural conclusions.

TABLE 4

Assignment of transitions by exact mass measurements on metastable ions

Transition	Calculated m^*	Measured m^*
$C_6H_7N^{+\cdot} \rightarrow C_5H_6^{+\cdot} + HCN$	46.876	
$C_6H_7N^{+\cdot} \rightarrow C_4H_4N^+ + C_2H_3^{\cdot}$	46.858	46.883
$C_6H_6N^+ \rightarrow C_5H_5^+ + HCN$	46.854	
$C_6H_6N^+ \rightarrow C_4H_3N^{+\cdot} + C_2H_3^{\cdot}$	45.936	45.958

Chapter 6

Approaches to the structures of gaseous ions

"This Inquiry, I must confess, is a gropeing in the Dark; but although I have not brought it into a clear light; yet, I can affirm that I have brought it from an utter darkness to a thin mist. . . "

John Aubrey, *Brief Lives*

INTRODUCTION

There follows an attempt to present the current status and to suggest some possible points of attack on one of the more challenging problems in science — determination of the structures of organic ions which survive for microseconds in the gas phase.

This is a field of recent development and it is therefore presented in its entirety, earlier work being dicussed briefly whenever there is overlap with extant treatments[76, 85, 146, 196]. Several methods other than those directly concerned with reactions of metastable ions are applicable to the ion structure problem and these are included in this discussion.

The remainder of this section is devoted to establishing ground rules for the discussion and to describing the major difficulties inherent in the problem. First, "ion structure" can refer to several levels of sophistication: (*i*) sequence of atoms, (*ii*)stereo-structure and (*iii*) electronic structure. Most of our interest is concentrated on larger ions for which detailed electronic structure assignments are unattainable. Some electronic structure assignments in simple ions have already been discussed in other chapters. In particular, the use of ion kinetic energy spectrometry in assigning electronic states and reaction mechanisms to the fragmentation of $H_2S^{+\cdot}$ is discussed in Chapter 4. The use of the charge-stripping procedure for determining the electronic states of rare gas ions has also been described in this chapter. "Structure", therefore, will most frequently apply to the bonding arrangement, although stereochemical structure is also becoming accessible by mass spectrometry. In some cases one has to settle for the crudest type of structural information, *viz.* that ions from two sources behave identically (or non-identically). Actually, it can often be arranged that the structure of one of the ions is fairly secure, so even these results can be valuable. It should be noted, too, that mixtures of isomeric structures may be involved in some instances.

A second fundamental point is that structure and internal energy are often difficult to extricate using available experimental techniques. In particular, ions of identical structure which differ in internal energy will generally fragment at different rates, show somewhat different kinetic energy releases, exhibit different isotope effects and undergo different amounts of labeled-atom scrambling. To distinguish between different structures and different energies in these cases is difficult but not impossible, as is shown below.

A third major factor is the possible change in structure with time: on formation, the structure of the ion may correspond to that of the neutral precursor. Subsequently, electronic and structural reorganization may occur. Structure then becomes a function of observation time and in particular, different answers may be obtained depending on whether fragmentation reactions occurring in the source or in the field-free region are monitored. A particularly direct demonstration of this effect has been reported for $C_4H_8^{+\cdot}$ ions formed with different internal energies by charge transfer[257]. At low internal energies, the ions formed from different isomers showed very similar fragmentation behavior, indicating that most reactions proceed from a common molecular ion. At higher energies, direct fragmentation without prior molecular rearrangement to a common structure becomes important.

Most of the methods discussed below involve structural inferences based on the behavior of fragmenting ions. The characteristics of non-fragmenting ions are much more difficult to monitor except that their heats of formation can be measured and they can be collisionally excited or otherwise made to take part in ion—molecule reactions. The information acquired on fragmenting ions is really relevant to the whole system $P^+ \rightleftharpoons$ transition state $\rightleftharpoons F^+ + N$. Only by using other information or assumptions does this reduce to evidence on the structure of P^+ (and/or F^+).

The above limitations, considerable as they are, have not prevented some impressive advances, although the difficulties in characterizing even such simple ions as $C_7H_7^+$ and $C_6H_6^{+\cdot}$ should not be underestimated. A discussion of the $C_6H_6^{+\cdot}$ structure, which provides a typical case of the procedure and methods available, appears later in this chapter.

ABUNDANCES OF METASTABLE IONS

The techniques discussed in this section tend to be relatively crude, but they have catalyzed much important work. The product of the reaction of a metastable ion is a daughter ion, distinguished from other daughter ions only by the longevity of its parent ion (with all the consequences which follow from this distinction). It is, therefore, natural that one of the earliest methods of characterizing metastable ions was directly analogous to the use of fragment ions to characterize molecular ions and, hence, to deduce molecular structures. Thus, the set of all reactions of a particular metastable ion constitutes a fingerprint of that ion. Since the reactions of metastable ions were originally detected by the corresponding low-abundance "diffuse" peaks in normal mass spectra, such fingerprinting consisted in recognizing the presence or absence of certain transitions (corresponding to diffuse, non-integral peaks in the mass spectrum). For example, the ion $C_6H_6^{+\cdot}$, as formed from benzene (by direct ionization) and from anisole (by a fragmentation route), might be the subject of interest. The reactions $C_6H_6^{+\cdot} \rightarrow C_6H_5^+ + H^\cdot$, $C_6H_6^{+\cdot} \rightarrow C_4H_4^{+\cdot} + C_2H_2$ and $C_6H_6^{+\cdot} \rightarrow C_3H_3^+ + C_3H_3^\cdot$ are given by metastable ions of both species. In this test, at least, the ions behave identically and the simplest conclusion is that they have identical structures. The test is not highly specific, however, and only by much more extensive testing, by the methods to be described below, could differences in the relative abundances and kinetic energy releases in the three different reactions of the metastable ions be related to differences in internal energy, mixtures of structurally isomeric ions, etc. An example[242] can also be cited where isomeric ions from two sources do not behave identically. $C_2H_5O^+$ metastable ions formed from primary alcohols, ethyl ethers and certain other compounds undergo loss of C_2H_2. Those formed from methyl ethers under the same conditions do not; hence, they probably have a different structure.

The above procedure does not do justice to the large amount of information actually resident in each metastable transition, information inherent in the variety of shapes of metastable peaks. The advent of more sensitive instruments for the examination of organic compounds revealed large numbers of metastable peaks in most spectra as well as characteristic shapes — "gaussian", "flat-topped", "dish-topped", etc. (These are discussed in detail in Chapter 3.) This means that the ion fingerprinting method could be extended to include both the occurrence of a particular reaction of a metastable ion and the shape of the resulting metastable peak. A landmark in the application of this procedure was Brown and Djerassi's study[71] on carbonates and the corresponding ethers formed by CO_2 loss. The metastable peaks in the spectra of methyl phenyl carbonate and anisole were found to be virtually superimposable. This is good evidence that the $C_7H_8O^{+\cdot}$ ions are very similar viz. identical in gross structure and similar in their internal energy distributions.

The first steps towards the quantitative use of the characteristics of metastable peaks in determining ion structures were made by Shannon and McLafferty[242] following methods initiated by Rosenstock et al.[234] and Hamill and coworkers[211]. The parent–daughter relationship, the essence of a metastable transition, suggested that daughter ions associated with different time scales be compared. Thus, if two ions have the same structures and internal energies, they should fragment to equal extents in any chosen time interval. Moreover, the abundance of the daughter ions formed in the source, expressed relative to the abundance of daughter ions formed in the field-free region, should be a parameter which characterizes the ion in question. Hence, the quotient $[m*]/[F^+]$, where square brackets represent abundances, came into use. An early application[188] of this method was to the loss of the substituent from isomeric meta- and para-substituted phenols. The observation of identical abundances and shapes of metastable peaks suggested that substituent randomization precedes fragmentation in the bromophenols, but large differences in the metastables indicated that this does not occur in the nitrophenols.

A major difficulty encountered in using $[m*]/[F^+]$ and the related parameter $[m*]/[P^+]$ is their sensitivity to ion internal energies. A study of the $C_{10}H_{12}O^{+\cdot}$ ion generated by ionization of diphenyl ether and by CO_2 loss from diphenyl carbonate, gave $[m*]/[P^+]$ ratios for the further fragmentation

$$C_{12}H_{10}O^{+\cdot} \rightarrow C_{11}H_{10}^{+\cdot} + CO$$

which differed by a factor of four[276]. This difference could be due to differences in the structures of the $C_{12}H_{10}O^{+\cdot}$ ions or to differences in their internal energies. The kinetic energy release accompanying this reaction is 435 meV when the reactant ion is generated by direct ionization and 438 meV when the carbonate is the precursor[155]. This result indicates that the reactive ion has a single structure but this information is not available from the metastable ion abundance results. These latter results, however, do emphasize the role of internal energy in metastable ion reactivity[277].

Empirically, it is observed that whenever the same ion is generated by a fragmentation route on the one hand and by direct ionization on the other, $[m*]/[P^+]$ values are greater for the former, indicating a higher average internal energy in the ions formed by fragmentation[276]. This has been explained as a consequence of the fact that ions formed by the fragmentation route may not be generated in their ground state but will frequently, even at threshold, possess excess internal energy. However, this excess energy is probably not the only factor involved since it often will not exceed the activation energy required for subsequent reaction. Another consideration is the time lag necessary to effect fragmentation. The field-free region is reached sooner after generation of the reactant ion in the fragmentation case and the sampling of higher-energy ions can account for the observed effects. In addition, if the $P(\epsilon)$ function is not a smooth curve, the ion generated by fragmentation could give greater or smaller metastable peak intensities.

With the emergence of a good understanding of the energy dependence of unimolecular rate constants, the nature of the internal energy distribution in gaseous ions, the kinetic shift and the whole of the QET as applied to organic ions, came such new advances as the detection, using the abundances of the metastable ion reaction products, of bond-forming fragmentations. Correlation of abundant metastable peaks with low activation energy reactions follows quite simply from QET. In addition, low frequency factor reactions tend to give abundant metastable ions[82, 152]. Add to this the recognition[273] that rearrangement reactions often have high entropies of activation and rather low activation energies, and one comes to the conclusion that $[m*]/[F^+]$ ratios for rearrangements must be generally larger than for simple cleavages[193]. While such information is relatively crude, it does provide an essential step in tracing ion structure from precursor to daughter ion and it illustrates a rather more general use of the quotient $[m*]/[F^+]$.

A method of using metastable ion abundance data for ion structure studies which has proved durable is the comparison of the relative abundances of ions due to competitive metastable transitions. The ratio $[m_1*]/[m_2*]$ where $[m_1*]$ and $[m_2*]$ are the abundances of the metastable ions, is far less sensitive to the internal energy of the ion than are the other ratios discussed above[242]. It can still show some variation[221], especially if the reactions involved have particularly low frequency factors, but the similarities are more often remarkable. Indeed, one might note that most of the recent applications of the method have found rather similar ratios, implying identical ion structures, for ions of a particular empirical formula generated from several sources [95], [247]. In a few cases where the ratios differed substantially, one metastable ion reaction was usually entirely absent[271], making it obvious that differences in ion structures were involved. One interesting case in which the abundance ratio showed an intermediate degree of variation is the case of $C_6H_6^{+\cdot}$ formed from benzene *versus* $C_6H_6^{+\cdot}$ formed from benzenechromium tricarbonyl. If the reactions of the metastable ions

$$C_6H_6^{+\cdot} \rightarrow C_4H_4^{+\cdot} + C_2H_2$$

$$C_6H_6^{+\cdot} \rightarrow C_3H_3^{+} + C_3H_3^{\cdot}$$

are associated with metastable peaks m_1^* and m_2^*, respectively, then the ratio $[m_1^*]/[m_2^*]$ is 2.86 for benzene but only 0.50 for the tricarbonyl[77]. It is not known whether different structures are involved or not. However, it seems probable that the change in the average internal energy of $C_6H_6^{+\cdot}$ ions generated as the final product of four successive fragmentation reactions will be sufficient to alter the abundance ratio. It should be noted, however, that $C_6H_6^{+\cdot}$ may exist in and fragment from isolated electronic states, it having been claimed that loss of C_2H_2 and loss of $H^{\cdot}$ are non-competitive reactions[4], [5].

One practical aspect of the use of the abundance ratio of competitive metastable ion fragmentations deserves attention. If the metastable peaks due to the two processes are very dissimilar in abundance, the quotient will be a large number and due to experimental error, differences in quotients for ions from different sources will be relatively large. The incorrect assignment of the product of the loss of two olefins from dialkyl ketones as the oxonium ion structure[191], [198] was due to this factor. It may also be noted that care must always be used, especially when metastable peaks of low intensity are examined, to ascertain whether or not they are truly the results of unimolecular reactions. If they are not, the collision-induced process is characteristic of the non-reacting ion, the unimolecular process is characteristic of the reacting ion, and these may not have the same structure. It is possible that the characterization of stable ions from two sources could be done on the basis of the relative abundances and occurrence of various collision-induced reactions. Much stricter control of the analyzer pressure is necessary in all such reactions than is required in the study of true metastable ion fragmentations.

ISOTOPIC LABELING

Another of the early techniques of ion structure elucidation involves the use of isotopic labeling. The use of isotopic labeling in determining molecular structures from mass spectra typically employs the assumption that gross changes in atom arrangement do not occur as the molecular ion fragments. This assumption becomes even more important when mass spectrometry is used to identify both the position and the amount of incorporated isotopic label. For example, in studies on the biosynthesis of gliotoxin, ^{15}N-glycine has been shown to be incorporated into both the indole and the N-methyl moieties[81].

The use of labeling to probe ion structures requires that the position of the label in the molecule be known; thereafter the arguments are similar to the above. For example, Schiff bases, Ar—CH=N—Ar' give prominent $(M-H)^+$ ions. Deuterium labeling of the methine hydrogen shows that the label is almost completely lost in the process leading to formation of $(M-H)^+$. Hence, a reasonable structure for the product ion is Ar—C≡N⁺—Ar'. By analogy, nitrones, Ar—CH = N⁺—Ar', and azines, Ar—CH = N—N = CH—Ar', which also give
$$\text{O}^-$$
prominent $(M-H)^+$ ions, might be expected to give analogous structures. However, the ring-deuteriated nitrone shown below undergoes deuterium loss[169] which indicates the cyclized structure for the $(M-H)^+$ ion in the case of nitrones.

Another more complex case in which isotopic labeling coupled with the study of metastable peaks enables one to suggest ion structures is the loss of water from benzyl benzoate. The results have been detailed in Chapter 5, but it is worth noting that in cases of this type, the evidence obtained is not so much about ion structures *per se*, rather it covers reaction mechanisms and hence, indirectly, ion structures.

A special aspect of the question of ion structure elucidation by isotopic labeling relates to the question of label scrambling which has considerable significance in regard to ion structure. Benzene is the best studied example of this type, the original observation[153, 181] that the hydrogens are randomized prior to fragmentations of metastable $C_6H_6^{+\cdot}$ ions having been proven[60, 99, 143, 223] to be due to both C—H bond cleavage and re-formation and to an independent carbon randomization reaction. It is this latter reaction which excites most interest since an attractive hypothesis which has photochemical analogies is that the $C_6H_6^{+\cdot}$ ion is in reversible equilibrium with one or more of its valence tautomers prior to fragmentation. Almost all the observations made to date on label randomization in aromatics (excluding toluene and other systems in which ring expansion occurs) can be explained by the occurrence of valence tautomerism with or without independent hydrogen migrations (but see the thiophene results below). Increased scrambling with increase in ion life-times can be detected by a study of the metastable peaks and indicates that the isomerization processes have very low activation energies and frequency factors. In some cases (e.g. C_2H_2 loss from ionized benzene) fragmentation may proceed from the valence tautomer (or some other isomer) rather than the initial species, but this has not been proven. Further discussion of the structure of the benzene molecular ion appears later in this chapter (p. 214).

The general features of the scrambling reactions which occur in the thiophen molecular ion are similar to those found in benzene. Carbon scrambling has been shown to occur in thiophen[96, 245] and in benzothiophen[86] while hydrogen randomization processes seem also to occur in both compounds. The migration of substituents about an aromatic nucleus by cleavage of the ring-atom substituent bond has been investigated in detail by studying metastable ion reactions in ^{13}C-labeled thiophens[227, 270]. It is found that the phenyl and bromo substituents, as least, can interchange positions on the skeleton with H atoms (and probably with each other). It also appears that these reactions occur after ring opening in the molecular ion as illustrated below. The radical site formed on ring opening can act as a receptor for the migration of such groups as H, Br and phenyl. The importance of radical sites in rearrangement mechanisms has been emphasized previously[196]

The high sensitivity of the IKES technique and related methods suggests a new method of ion structure determination based on the natural isotopic label of the ion. In favorable cases, reactions of the $(M+3)^+$ ion can be monitored and the isotopic distributions pertaining to these reactions provide information on the nature of the neutral fragment and, hence, on the reactant ion structure[64].

It is worth noting that the MIKES technique is particularly suited to studies on metastable ion abundances of the type discussed in this and the preceding section. The recent introduction of MIKE spectrometers should facilitate these studies and should be particularly useful when partially labeled compounds must be employed.

ION KINETIC ENERGY SPECTROMETRY

One property of metastable ions which has exceptional value as a means of characterizing gaseous ions is the kinetic energy, T, released in the fragmentation. Because of the close association which has developed between such measurements and the IKES technique, the whole topic is presented under the present heading. Collision-induced decompositions are also included; the kinetic energies of the products of such reactions reflect both the kinetic energy release, T, and the kinetic energy, Q', lost by the reactant ion.

As emphasized in Chapter 4, the selection by the instrument of a narrow range of internal energies of metastable ions means that the kinetic energy released in the fragmentation of a metastable ion is a characteristic of ion structure rather than of internal energy. To add just one more illustration here to those already given in Chapter 4, both anisole and nitrobenzene react to give an ion $C_6H_5O^+$, anisole by loss of $CH_3^{\,\cdot}$ from the molecular ion in what is believed to be a simple cleavage reaction[72, 89] and nitrobenzene by loss of $NO^{\,\cdot}$. The nitrobenzene reaction can occur by two mechanisms, a three-centered cyclic rearrangement to give the phenoxyl cation directly and a four-centered cyclic rearrangement to give a product which is expected to isomerize readily, after the fragmentation and conversion of internal to kinetic energy is complete, to the more stable phenoxyl cation by transfer of a hydrogen atom. Hence, the reactive $C_6H_5O^+$ ions from nitrobenzene and anisole are expected to be identical in structure (phenoxyl cations) though not in energy. The loss of CO from the metastable $C_6H_5O^+$ ions generated from anisole is accompanied by an energy release of 47 meV, while the corresponding ions generated from nitrobenzene and selected, because they are also metastable, to have a similar range of internal energies, release 48 meV (ref. 155). This selection overcomes differences due to the nitrobenzene molecular ion fragmenting by two mechanisms.

The comparison of kinetic energy releases provides a simple and reliable method for establishing the identity or non-identity of ions generated from different sources. It should again be emphasized that all the information obtained refers to ions that undergo fragmentation. No information .can be obtained by these methods about non-reacting ions. In order to obtain more detailed information on ion structures, situations in which $T \cong T^e$ (see p. 104) can be examined. These fall into two major classes, first the charge-separation fragmentations of doubly charged ions, $AB^{2+} \rightarrow A^+ + B^+$, and second, reactions of singly charged ions which have large reverse activation energies.

The charge-separation reactions of doubly charged ions can readily be detected by the metastable peaks resulting from fragmentations occurring in the field-free regions. The coulombic energy gained on separating the charges from their distance in the activated complex to infinity will be large. This accounts for the fact that such reactions are always accompanied by large kinetic energy releases ($T = 1-4$ eV). A large reverse activation energy ϵ_0^r is a necessary but insufficient condition for a large T, which requires, in addition, that a high proportion of ϵ_0^r be partitioned into kinetic energy. For charge separation, the assumption that all the coulombic energy appears as kinetic energy is justified by numerous measurements of kinetic energy release and their correlation with molecular structure[17], although it cannot be directly demonstrated. While there may be small contributions to the measured energy release in charge-separation reactions due to the internal energy of the activated complex or to the non-coulombic portion of the reverse activation energy, these contributions will be negligible relative to the coulombic contribution. Hence, the approximation $T = T_{coulombic}$ will seldom be in error by more than $0.1-0.2$ eV and if, as the available evidence suggests, many charge-separating reactions occur by simple cleavage, it will usually be in error by even less than this.

These considerations allow one to write for charge-separation reactions,

$$T = T_{coulombic} = \frac{e^2}{R}$$

where R is the interchange distance in the transition state. Expressing this in common units

$$T(eV) = \frac{14.39}{R(\text{Å})} \tag{45}$$

Because of the possibility of contributions to T due to factors other than the coulombic repulsion, the intercharge distance arrived at from eqn. (45) will represent a minimum value.

The calculated intercharge distance in the transition state helps to define the transition state geometry. Of course, bond lengths in gaseous ions are not accurately known, but there is considerable variation in the range of observed R values and crude structural distinctions are possible. They include assignment of the location of charges at the ends or nearer the center of a linear ion, determination of whether or not a cyclic ion undergoes ring-opening and determination of whether or not isomeric ions fragment from a common structure. This approach was first used[37] in assigning a linear structure to doubly charged benzene molecular ions based on the fact that the reaction

$$C_6H_6^{2+} \rightarrow C_5H_3^+ + CH_3^+$$

involves release of 2.6 eV. This corresponds to an intercharge distance of 5.5Å, which is reasonably associated with terminal charge sites in an acyclic doubly charged ion.

Loss of a methyl cation also occurs from the molecular ions of toluene and the xylenes. The kinetic energy releases measured from the metastable peak widths in these cases correspond to intercharge distances of 5.3Å for toluene and 5.6Å for each of the xylenes[7]. This result indicates that the charges in the reacting ions are separated by the same number of bonds and, hence, that the extra carbon atoms in the toluene and xylene ions must exist as branches on the main C_6 skeleton.

The evidence for ring opening in the above reactions can be compared with the behavior of the $(M-2)^{2+}$ and $(M-4)^{2+}$ ions formed from any of the xylenes[7]. For p-xylene, the following results are obtained

$$C_8H_8^{2+} \rightarrow C_7H_5^+ + CH_3^+ \quad T = 2.6 \text{ eV} \quad R = 5.6 \text{ Å}$$
$$C_8H_6^{2+} \rightarrow C_7H_3^+ + CH_3^+ \quad T = 2.6 \text{ eV} \quad R = 5.5 \text{ Å}$$
$$C_8H_4^{2+} \rightarrow C_7H^+ + CH_3^+ \quad T = 1.9 \text{ eV} \quad R = 7.7 \text{ Å}$$

While the molecular ion and the $(M-2)^{2+}$ ions apparently have analogous structures, a sharp difference in structure occurs in the $(M-4)^{2+}$ ion. The large intercharge distance (7.7Å) suggests that this ion has an unbranched linear C_8 structure; the formulae of the fragments suggest that the terminal carbon atoms carry three and one hydrogens, respectively.

The fact that o-, m- and p-xylene behave identically in the foregoing charge-separation reactions suggests that the doubly charged ions isomerize to common structures. This would parallel the behavior of the corresponding singly charged ions. Another important parallel, which may be due to the possibility of ring expansion in both singly and doubly charged ions, is found in the fact that complete hydrogen randomization precedes the reactions of metastable doubly charged molecular ions of toluene[9]. Moreover, the fast source reactions of this ion are accompanied by considerable scrambling prior to H_2 loss, but relatively little scrambling precedes H˙ loss[8, 204].

The doubly charged ion mass spectra ($2E$ spectra) and associated fragmentation patterns of hydrocarbons have been studied in detail[10, 55]. There is a striking tendency for formation of particular stable ions which operates in spite of gross differences in molecular structure. Tentative structures for these ions have been assigned on the basis of this evidence and the energy released in their charge-separating reactions.

The accumulated results of the hydrocarbon ion studies just described suggest that an average value of 1.2—1.3Å should represent one bond length in doubly charged ions (some variation with degree of unsaturation is, of course, recognized). This allows one to determine the number of bonds between the charges in reacting ions. This principle is valuable when the behavior of nitrogen-containing compounds is considered; here stability considerations which lead one to expect the charges to be well separated may be in competition with a strong tendency for charges to be localized on the heteroatoms. For example, the reaction $M^{2+} \rightarrow (M-28)^+ + 28^+$ occurs with an intercharge distance of between 5.0 and 5.5Å for p-phenylenediamine and for its mono- and dimethylated homologs. On the other hand, the same reaction in aniline, o-toluidine and 2,4-dimethylaniline shows intercharge distances of 3.6, 4.5 and 5.6Å (ref. 11). This increase is consistent with an increase of one bond length in intercharge distance in each step and indicates that one charge is localized on the heteroatom while the other takes a remote position. The tendency for heteroatoms to bear the charges in doubly charged ions is also shown by the way in which such ions lose neutral fragments. The loss of HCN is a favored mode of fragmentation of singly charged ions containing nitrogen atoms and it also occurs for doubly charged ions which have three nitrogen atoms. However, if the doubly charged ion has only two nitrogen atoms, C_2H_2 loss is observed instead of HCN loss, implying that nitrogen can now only be lost in charged fragments because the charges are localized on the two nitrogen atoms.

One complication in the consideration of ion structures, which has already been mentioned in connection with singly charged ions and which is very evident with doubly charged ions, is the fact that an ion may exist in two or more different structures corresponding to different modes of fragmentation. This is seen, for example, in the loss of H_2CN^+ ($R = 3.6$Å) and $C_2H_3^+$ ($R = 6.0$Å) from the doubly charged molecular ion of aniline. By way of contrast, the same reactions in 2,3-dimethylaniline both have intercharge distances of 5.7Å. o-Phenylenediamine provides a further example: the loss of 17^+ from the molecular ion has a corresponding intercharge distance of 6.6Å, while the loss of 18^+ shows the unusually large value of 11.1Å.

A final observation on kinetic energy release in the fragmentation of doubly charged ions provides some indication of bond lengths in the transition state. The ion $(C_4H_2N_2)^{2+}$, loses H^+ giving a calculated intercharge distance of $\sim$11Å in pyrazine, $\sim$10.5Å in pyrimidine and $\sim$9.5Å in pyridazine. These results imply a C—H bond distance of 2—3Å in the transition state from which H^+ is lost. The corresponding C—D distance is 0.5—1.0Å greater.

In summary, the presence of two charges makes it possible to "tag" two atoms and examine their relative positions in the fragmenting doubly charged ion. Studies on substituent isomerization and ring-opening in aromatic ions are greatly facilitated. In fact, it may eventually prove a simpler proposition to define the structures of fragmenting doubly charged rather than singly charged ions. Already, in work of the type described above, guidelines are emerging for assigning doubly charged ion structures: values of kinetic energy released, the energetic advantanges of maximum charge separation permitted by the bonding arrangement, isotopic labeling data and the importance of product ion stability are all taken into account. The tendency for charges to be localized on the most favorable sites, particularly heteroatoms, also is evident.

The other major application of the $T \cong T^e$ relationship is to fragmentations of singly charged ions which involve bond formation as well as bond cleavage. One example of the use of kinetic energy release measurements in conjunction with thermochemical data to deduce ion structures and reaction mechanisms has already been given in connection with the discussion of energy partitioning in Chapter 4. The example discussed there concerns the loss of formaldehyde from anisole and substituted anisoles. Below, several other reactions are considered, including NO˙ loss from nitrobenzene and substituted nitrobenzenes and HCN loss from benzaldoxime O-methyl ethers. An attempt is also made to draw generalizations from all the available energy partitioning studies on organic ions, although it is recognized that the limited data make the conclusions tentative only.

One of the earliest applications of the use of kinetic energy release in an ion structural problem was the study of $NO^{\bullet}$ loss from nitrophenols[41, 75]. The p-nitrophenol ion, whether formed by direct ionization or from the corresponding phenetole, undergoes $NO^{\bullet}$ loss with release of considerable kinetic energy, but the corresponding meta-substituted compounds show gaussian metastable peaks, i.e. small values of energy release. Apparently, ϵ_0^r is large in the para case but small in the meta case. The simplest interpretation of this result is that the para compound generates the stable quinonoid ion[41] while the meta compound generates a product ion of lower stability. This result establishes that (i) the molecular ions do not isomerize prior to $NO^{\bullet}$ loss, (ii) oxygen transfer occurs to the originally substituted ring position. These significant ion structural observations are illustrated below.

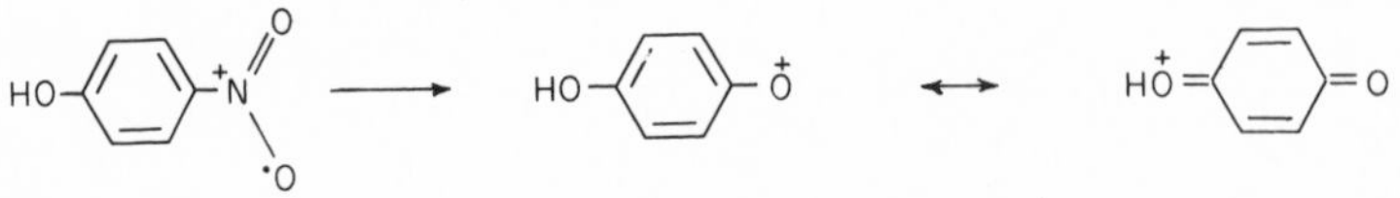

These and other results led to the conclusion that $NO^{\bullet}$ loss   from nitrobenzenes occurs by the three-membered cyclic rearrangement shown.

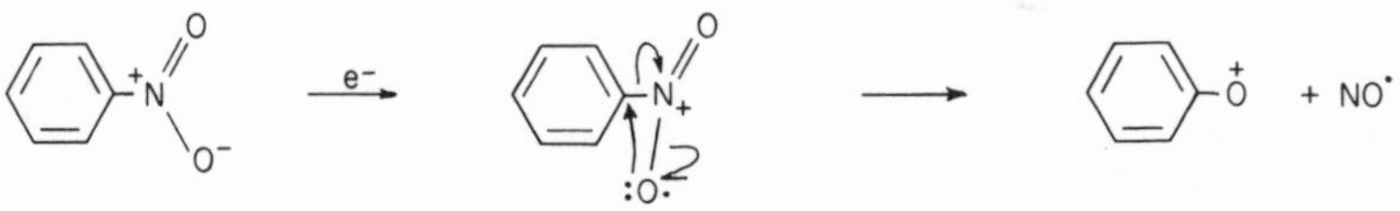

However, if the metastable peaks due to these reactions are examined under conditions of high energy resolution, they are usually observed to be composite. The peak for p-chloronitrobenzene is shown in Fig. 92. Deconvolution shows that the process giving the larger energy release ($T_1 = 0.83$ eV) makes up approximately 60% of the signal while the smaller energy release ($T_s = 0.11$ eV) makes up 40%. Neither process is collision-induced. The corresponding data for nitrobenzene itself are $T_1 = 0.55$ eV (50%) and $T_s = 0.09$ eV (50%). These observations mean that there must be a second mechanism

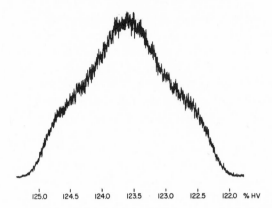

125.0   124.5   124.0   123.5   123.0   122.5   122.0 % HV

Fig. 92. Composite metastable peak for loss of NO˙ from the molecular ion of
p-chloronitrobenzene. The reaction takes place by two competitive mechanisms.

whereby nitrobenzenes lose NO˙. This is associated with a much
smaller energy release than is the first reaction and it becomes the
dominant process as the *para*-substituent is made more electron-
withdrawing. For example, the p-cyano derivative releases 0.35 eV
and 0.07 eV, the processes comprising 40% and 60%, respectively, of
the total signal. On the other hand, the p-amino compound shows
>95% contribution from the process releasing the larger amount of
energy (1.24 eV). The kinetic energy release T_s is approximately
independent of substituent. This is in contrast to the effect of
changing the substituent upon T_1 and suggests that this mechanism
involves oxygen transfer via a four-membered cyclic transition state
to the *ortho* position. Further evidence for this conclusion is presented in the original paper[62].

The ionization potential of the phenoxyl radical has been mea-
sured, so that it is possible to calculate $\epsilon_{excess} = (\epsilon_0^r + \epsilon^{\ddagger})$ for NO˙
loss from nitrobenzene via the three-membered cyclic transition
state. The energy partitioning quotient T/ϵ_{excess} for this reaction is
0.55/0.79 = 0.70. That is, a large fraction of the available energy
appears as kinetic energy in this low-entropy rearrangement. There is
indirect evidence[62] that this is also true of the corresponding reactions of the substituted nitrobenzenes. Now, since the measured
energy release is itself a large fraction of the available energy and
since it represents only an average value, the maximum release, as
measured from the metastable peak width at the base line, will even
more closely approximate ϵ_{excess} and can be used to estimate the
heats of formation of the substituted phenoxy cations. This in turn
allows one to calculate the ionization potentials of the substituted

phenoxyl radicals since the heats of formation of the neutral radicals are known. These results, plotted against σ^+ (with which IP's should correlate) are given in Fig. 93. It can be seen that a good linear correlation results, showing, at least, the internal consistency of the method (independent values are not available for comparison).

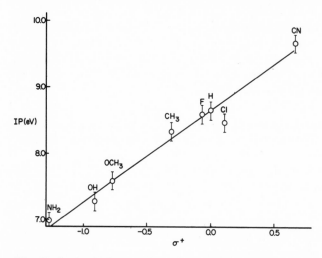

Fig. 93. Ionization potentials of *para*-substituted phenoxyl radicals calculated by the energy partitioning method and plotted against σ^+.

HCN elimination from benzaldoxime O-methyl ethers involves methoxyl rearrangement to the aromatic ring and could occur via either a four- or a five-membered cyclic transition state[63] as shown below.

Metastable molecular ions give energy releases, when they dissociate, which are at a maximum for strongly electron-donating and strongly electron-withdrawing substituents. Thus, the *p*-methoxy compound releases 0.67 eV, the p-nitro 0.62 eV but the unsubstituted compound only 0.37 eV. If ϵ_{excess} is calculated on the assumption that the corresponding ionized anisole is generated, T/ϵ_{excess} is 0.73 for the p-methoxy compound, 0.39 for the p-nitro compound and ∿0.22 for most other substituents. Although the metastable peak shapes did not show evidence for a composite structure, the variation in T/ϵ_{excess} with substituent suggests that this is present and that both types of reaction occur in these ions. The high value of the energy-partitioning quotient for the p-methoxy compound is consistent with reaction largely by the four-membered cyclic rearrangement in this case. The particular stability of the substituted anisole product ion in this case is in agreement with this interpretation.

A $para$-substituted oxime ether will yield a $para$-substituted anisole if it reacts by the first mechanism and it will yield a $meta$-substituted anisole (after isomerization of the ionized carbene) if it reacts by the second. Now it is known that $meta$- and $para$-substituted anisoles behave differently as regards reactions of their metastable ions. In particular, H_2CO loss gives a composite metastable peak and the relative proportions of the two components as well as the individual energy releases are different for the two isomers. These facts suggest that the mechanism by which oxime ethers lose HCN in the ion source can be determined by examination of the metastable peak shapes for further loss of H_2CO at high energy resolution. The application of this method to p-chlorobenzaldoxime-O-methyl ether is shown in Fig. 94 which juxtaposes the metastable peaks for the reactions.

(a) $(M-HCN)^{+\cdot} \rightarrow (M-HCN-H_2CO)^{+\cdot}$ in the oxime ether
(b) $M^{+\cdot} \rightarrow (M-H_2CO)^{+\cdot}$ in p-chloroanisole
(c) $M^{+\cdot} \rightarrow (M-H_2CO)^{+\cdot}$ in m-chloroanisole.

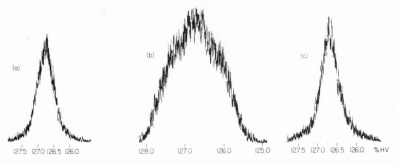

Fig. 94. Metastable peaks obtained by the HV scan method for the reaction $142^+ \rightarrow 112^+ + 30$ in (a) p-chlorobenzaldoxime-O-methyl ether, (b) p-chloroanisole and (c) m-chloroanisole.

The metastable peaks for (a) and (c) are identical in shape, proving that the reaction occuring in the ion source involves the five-centered methoxyl rearrangement. This mechanism apparently also explains the fragmentation process for low-energy metastable molecular ions as already discussed.

The technique just presented could find general application in the problem of determining the mechanisms by which high-energy ions undergo rearrangment reactions (other than hydrogen rearrangements). The use of reactions of metastable ions as the probe means that suitable further reactions can almost always be found. The metastable peaks provide a specific probe for a particular ion and both abundances and detailed peak shapes characterize the *preceding* ion source reaction.

This oxime ether study also highlights another general feature of competitive reactions of metastable ions, the tendency for activation energy to control reactivity in low-energy ions but for frequency factors to be dominant for high-energy ions. Thus, the *p*-methoxy oxime ether undergoes both the four- and and five-centered rearrangements in the ion source, but largely the four-centered reaction when low-energy metastable ions are sampled.

Using the methods outlined above and those described in connection with the fragmentation of substituted anisoles in Chapter 4, energy-partitioning data has been obtained on several other compounds and series of compounds. In each case the energy partitioned from the reverse activation energy is studied. *n*-Butylbenzene loses C_3H_6 by a mechanism which ICR evidence[7][8] suggests involves six-rather than four-centered hydrogen transfer. The energy-partitioning method gives a quotient of 0.024 calculated on this basis[9][1] and this small value is apparently typical of a mechanism involving a large ring in the transition state. Benzyl methyl ether loses formaldehyde by a process which might involve a four-membered hydrogen transfer. The energy-partitioning parameter calculated for this reaction is 0.59. This high value is reasonable for the proposed mechanism. Finally, a series of substituted phenyl ethyl ethers (phenetoles) and acetanilides has been studied.

The process of interest is the loss of $\dot{C}_2H_4$ from the phenetole molecular ion and the loss of H_2CO from the acetanilide molecular ion. The major proposed mechanisms are, respectively, hydrogen transfer through a four-membered cyclic intermediate to give the ionized phenol or aniline and hydrogen transfer through a six-membered cyclic intermediate to give the ketonic or imino ion. The two possible mechanisms in the case of the phenetole are illustrated below.

Almost all the methods available to the organic mass spectroscopist have been used to tackle this problem. Among them, the results from two, both of which are applicable to low-energy ions, may be noted. First, the use of thermochemical measurements suggested the six-membered cyclic transition state while a k_H/k_D isotope effect study on metastable ions suggested the alternative 1,3-hydrogen tranfer.

All the *meta*- and *para*-substituted phenetoles studied showed non-composite metastable peak and all gave rather small T values. Now, it is possible to estimate the reverse activation energy for each of the two proposed reactions using published thermochemical measurements and group equivalent arguments. For phenetole itself, one finds $\epsilon_{excess} = \epsilon_0{}^r + \epsilon^{\ddagger} = 1.94$ eV for phenol formation and 0.83 eV for formation of the cyclohexadienone. Now, if $\epsilon^{\ddagger}$ is negligible and reaction occurs by the four-membered cyclic transition state, $\sim1\%$ of the reverse activation energy appears as kinetic energy. If the six-membered cyclic intermediate is involved, the value is 3%. Similar data were found for acetanilide. The relationship between T and $\epsilon_0{}^r$ is not yet known, in general, but sufficient evidence is available to indicate that the value of $\sim1\%$ is unreasonable for a concerted four-membered cyclic rearrangement. A step-wise process is possible as is a concerted six-membered cyclic rearrangement. The continued acquisition of data on other energy-partitioning reactions should soon allow a completely unequivocal answer to this and to other such mechanistic questions.

In attempting to summarize the results so far obtained on energy partitioning in organic ions, it must first be noted that the available thermochemical data are not particularly accurate. However, within any particular series of substituted compounds they are consistent and it is found (for nitrobenzenes, anisoles and phenetoles) that the energy-partitioning quotient is essentially constant. When it is not (as for oxime ethers) there is independent evidence for two contributing unresolved processes. The generalization that large quotients are associated with concerted reactions occurring through cyclic activated complexes of small ring size seems secure. The three-membered cyclic rearrangement in the nitrobenzenes gave an average value of ~ 0.63, the four-membered cyclic rearrangement in anisoles ~ 0.16 and the six-membered cyclic rearrangement in n-butylbenzene and other compounds ~ 0.03. However, little reliance can be placed on actual values because of (i) approximations in the treatment, $e.g.$ $\epsilon_{excess} \cong \epsilon_0{}^r$, ($ii$) the complex shape of the potential energy surface cannot be treated and the question of whether the activated complex comes "early" or "late" has been ignored. Hence, it is not surprising that the four-centered reaction in p-methoxybenzaldoxime ether has a quotient as high as 0.73. On the other hand, within this series of compounds the competitive five-centered rearrangements give much lower quotients, as expected.

Only a very limited amount of data on energy partitioning in simple bond cleavages is yet available. One case which has been studied is methyl radical loss from anisoles. This example fulfills the requirement that ϵ_{excess} be large and $\epsilon^{\ddagger}$ small, so that T/ϵ_{excess} can be equated to $T^e/\epsilon_0{}^r$. The measured quotient for this reaction is small, about 0.1, for all substituents. However, the methyl-substituted anisoles showed composite peaks for methyl radical loss and the new process involved a large value of T (ref. 89). It is concluded that in these cases a competitive rearrangement reaction occurs, probably by ring expansion with formation ultimately, of the stable protonated tropone species as the product ion.

A problem which has to be faced in interpreting energy-partitioning quotients is the possibility that rearrangement will occur in the molecular ion and that fragmentation, the final step in the potential energy profile and the one which determines the energy partitioning, will occur by a simple bond cleavage. Such a step-wise reaction will be indistinguishable from a concerted process, but it will involve a "looser" activated complex and will therefore give a smaller T/ϵ_{excess} value than expected.

Although the bulk of this chapter is devoted to the elucidation of gross ion structures, it is worth mentioning that measurement of T has been applied successfully to the determination of the electronic states involved in the fragmentation of small ions. The discussion of $H_3{}^+$, $H_2S^{+\cdot}$ and other simple ions in Chapter 4 covers structural as well as mechanistic questions and will not, therefore, be repeated here. The discussion of $H_2O^{+\cdot}$ later in this chapter (p. 205) and in Appendix I is also relevant.

ISOTOPE EFFECTS

Isotope effects both upon abundances of metastable ions and upon the kinetic energy released in the reactions of metastable ions have been used to gather information on ion structures and on reaction mechanisms. The theory associated with these applications has been treated briefly in Chapter 4. Although our understanding of these effects is by no means complete, clear structural information has emerged in some cases. For example, all available data indicates that secondary isotope effects have only a small effect upon T. The loss of N_2 from *sym*-triazole gave a T value of 1.41 eV while the d_1-analog gave 1.57 eV (ref. 69). This relatively large difference is probably not due to a secondary isotope effect and Shannon concludes that hydrogen migration occurs in the transition state.

The use of primary isotope effects, in particular k_H/k_D and T_H/T_D, as a means of acquiring information on ionic reaction mechanisms has been met in Chapter 4. Values of T_H/T_D substantially less than unity are indicative of a tunneling mechanism for $H^\cdot (D^\cdot)$ loss. The magnitude of k_H/k_D, on the other hand, provides some information on the rate of increase of the rate constant, k, with ion internal energy.

Toluene provides an interesting case in which k_H/k_D values have proved of considerable value in deducing ion structure. For ions which react in the ion chamber, k_H/k_D is 1.58 and there is a preference factor of 1.32 for loss of a methyl hydrogen over loss of a ring hydrogen[201]. On the other hand, metastable ions show k_H/k_D = 2.40 but no preference factor[31, 147]. The increased isotope effect in low-energy ions is in agreement with expectation. The preference factor seen in high-energy ions is larger than that which favors $H^\cdot$ loss from the 7-position in cycloheptatriene and indicates that toluene fragments both by direct $H^\cdot$ loss from the methyl group as well as by isomerization to cycloheptatriene at high energy[147]. It is interesting to note that doubly charged toluene molecular ions which fragment by $H^\cdot$ loss in the ion source do so with a k_H/k_D isotope effect of 1.3 and a preference factor of 3.3 in favor of methyl hydrogen loss[8]. This much higher preference factor is consistent with formation of a linear ion and with relatively slow hydrogen randomization within the ion source residence time. The slow randomization indicates that the mechanisms by which this can occur probably are different to those in (cyclic) singly charged toluene molecular ions. The lower average internal energy of the doubly charged toluene molecular ions effectively depresses the preference factor which remains high even in spite of this. The conclusion that linear ions are involved is therefore strengthened.

Deuterium isotope effects upon metastable ion abundances have been employed to determine reaction mechanisms by a method[148, 261] which can be illustrated for the reaction

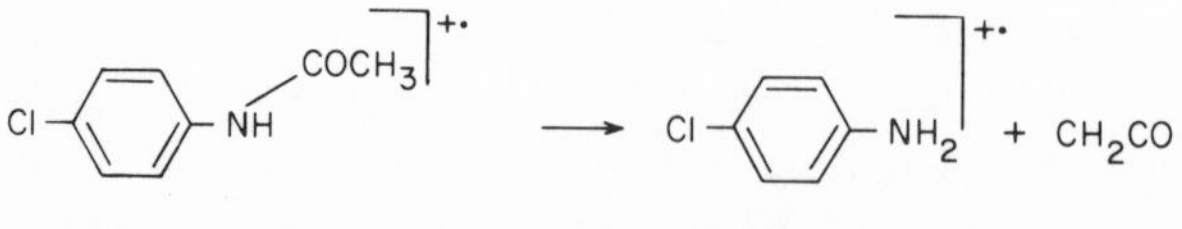

This reaction may occur by hydrogen transfer via a four-membered cyclic transition state to give the ionized aniline directly or it could involve a six-membered cyclic intermediate which would yield the iminocyclohexadiene shown below. Subsequent isomerization of this ion to give the aniline might precede its further fragmentation.

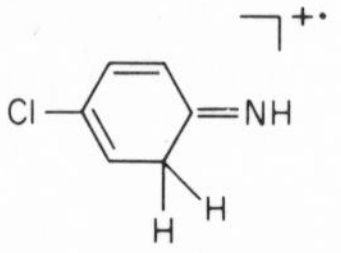

Now, aniline itself fragments by loss of HCN apparently after isomerization to the imino structure. Hence, the molecular ion of p-chloroaniline undergoes the sequence

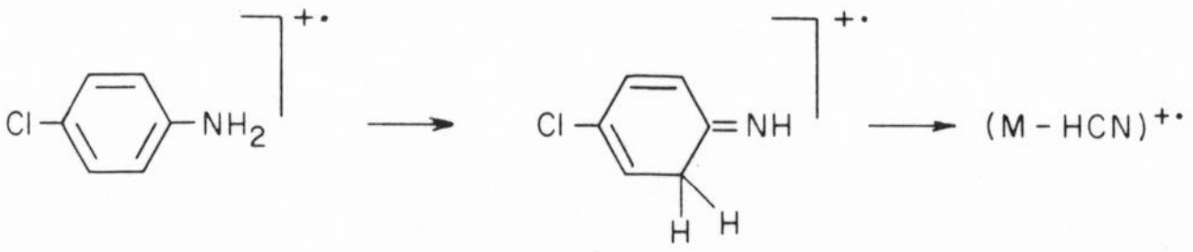

This molecular ion shows metastable peaks for both Cl$^{\cdot}$ loss and for HCN loss. If these processes are competitive and if, as is reasonable, there is no primary isotope effect associated with Cl$^{\cdot}$ loss, then the ratio of these metastable ion abundances can be used to determine whether an isotope effect operates for HCN elimination. This was found to be the case. Thus, the abundance ratio was about 30% lower in N,N-d$_2$-aniline than in aniline itself. Now, if the p-chloroaniline molecular ion were generated from the acetanilide and its further fragmentation studied, a similar isotope effect should occur. By way of contrast, if the loss of ketene occurs by a six-membered cyclic rearrangement, the iminocyclohexadiene would be generated directly and HCN loss would not be expected to exhibit an isotope effect as measured relative to the reference process, Cl$^{\cdot}$ loss.

An isotope effect was observed, and it was similar in magnitude to that found for p-chloroaniline itself. It is, therefore, concluded that the four-membered cyclic rearrangement occurs.

THERMOCHEMISTRY

The assignment of ion structures on the basis of thermochemical measurements is a well-developed subject which has been reviewed recently[113, 129]. The method has the advantage that it provides data on ions at threshold, that is, on non-reacting ions. There are, however, two major difficulties involved in this method of characterizing ions: the first is the frequently poor quality of the ionization and appearance potential data on which calculated enthalpies of gaseous ions depend. This problem is, however, amenable to solution. The second poses more fundamental difficulties: there is at present no satisfactory way of correcting for the excess energy terms involved in appearance potential determinations.

The excess energy terms comprise the kinetic shift and the reverse activation energy (Chapter 4). In addition, the Franck—Condon factor may cause the minimum internal energy which can be deposited in the ion to exceed that required to satisfy the kinetic shift.

In simple cleavage reactions, the reverse activation energy is believed to be small or even negligible (there is little direct evidence for this latter assertion, although experimental data on ion-radical reactions suggest that small activation energies are common). The differences in heats of formation of isomeric compounds are frequently large enough to make it possible to assign structures from uncorrected appearance potential data in the case of simple cleavage reactions. In fact, the heats of formation values for ions formed from a variety of sources often cluster around a small number of discrete values which are therefore characteristic of a corresponding number of isomeric structures.

The $C_6H_5^+$ ion, for example, has been considered in this way without correction for excess energy terms. A value for the ionization potential of the phenyl radical is available and leads to a ΔH_f value of 285 kcal. $mole^{-1}$, appropriate to the cyclic ion. Phenyl bromide yields a similar value (288 kcal. $mole^{-1}$) and is, therefore, assumed to yield a cyclic $C_6H_5^+$ ion. On the other hand 3,5-hexadien-1-yne has an uncorrected heat of formation of 348 kcal. $mole^{-1}$. This is reasonable evidence that the ion formed in this way is structurally dissimilar to that generated by ionization of the phenyl radical.

The method of Franklin and coworkers for correcting measured appearance potentials for the excess energy term associated with the internal energy of the activated complex has already been discussed in Chapter 4. Their empirical relationship between the measured kinetic energy released in a fragmentation and the associated excess energy of the activated complex was derived from reactions of known thermochemistry. Application of this relationship to reactions of unknown thermochemistry generally yielded enthalpies of ions which were in reasonable agreement with all available data. In several cases decisions regarding ion structure based on the refined thermochemical data obtained by Franklin's method have been made[115, 127] although some reservation has been expressed as to the accuracy with which the excess energy term is determined in the time of flight procedure.

Franklin's procedure cannot be used for elimination and rearrangement reactions which normally have appreciable reverse activation energies. In such cases, the reverse activation energy will normally be the major excess energy term. It has already been noted in Chapter 4 that for this type of reaction, the kinetic energy released by metastable ions will be largely due to partitioning of the reverse activation energy. Hence, correction of appearance potential data for the excess energy term requires that the relationship between T and ϵ_0^r be known. Progress is being made in this endeavor, as described earlier in this chapter.

As an illustration of the magnitude that ϵ_{excess} can attain in some reactions, consider formaldehyde elimination from anisole.

$$C_6H_5OCH_3^{+\cdot} \rightarrow C_6H_6^{+\cdot} + CH_2O$$

The available appearance potential values for $C_6H_6^{+\cdot}$ are 11.50 eV and 11.27 eV, averaging at 11.39 eV. Hence

$$11.39 \text{ eV} = \Delta H_f(C_6H_6^{+\cdot}) + \Delta H_f(CH_2O) - \Delta H_f(C_6H_5OCH_3) + \epsilon_{excess}$$

There is little uncertainty in the heats of formation of the neutral species: -28 kcal.mole^{-1} for CH_2O and -16 kcal.mole^{-1} for anisole. Using these values and ignoring ϵ_{excess}, one finds

$$\Delta H_f(C_6H_6^{+\cdot}) = 275 \text{ kcal mole}^{-1}$$

However, the heat of formation of the benzene molecular ion, which is believed to be formed in this case, is 233 kcal. mole^{-1}. Hence, the excess energy term is estimated to be 42 kcal .mole^{-1}.

Measurements of appearance potentials to obtain thermochemical data as ion enthalpies have been made since the earliest days of mass spectrometry. In retrospect, much of this work can be seen to be of limited use because of insufficient attention being given to correcting for the excess energy of the activated complex and for the reverse activation energy of the reaction. Recent work of Lossing using mono-energetic electrons shows that measurements can be carried out to the requisite degree of accuracy. His reactions involved simple cleavage of C—H bonds in which reverse activation energy is expected to be small. Nonetheless, when the translational energy release during the reactions was measured using IKES, an enthalpy correction of at least 3 kcal/mole was shown to be necessary.

A new thermochemical method can be suggested for characterizing non-reactive ions. As detailed in Chapter 4, singly charged ions undergo collision-induced ionization with the loss of some kinetic energy, Q'. From the measurement of this kinetic energy, one gets the difference between the double and single ionization potentials of the ion. This quantity can be measured accurately to about 0.5 eV and this should be sufficient to distinguish isomeric stable ions in at least some cases.

The rather different questions one meets in applying thermochemical methods to the structures of small ions is illustrated by the situation for the reaction $H_2O^{+\cdot} \rightarrow O^{+\cdot} + H_2$. The structural question here concerns the electronic structure of the product ion $O^{+\cdot}$. There is no reaction of metastable ions leading to $O^{+\cdot}$ formation from water and a variety of evidence indicates that the ion source reaction involves predissociation[111]. The appearance potential of $O^{+\cdot}$ is high ($\sim$ 18.8 eV) and the ionization efficiency curve indicates a low probability for formation of ground-state products and a much higher probability for formation of products in an excited state ($\sim$ 28 eV)[174]. It may also be noted that the alternative mode of fragmentation of $H_2O^{+\cdot}$ to give $HO^+ + H^{\cdot}$ also occurs by predissociation and has a lower appearance potential. This accounts for the low abundance of $O^{+\cdot}$ in the mass spectrum of water and for the absence of a metastable peak corresponding to slow predissociation as is observed for the analogous reaction in $H_2S^{+\cdot}$ (see Chapter 4).

Upon collision, the transitions of stable $H_2O^{+\cdot}$ ions, which exist predominantly in the low vibrational states of the ground electronic state, are approximately those induced by photon or electron impact[157]. These collisions occur with the conversion of 22 ± 4 eV (the range covering different collision gases) of kinetic energy into internal energy and with the subsequent liberation of 1.3 eV of energy as translational energy of the products, $O^{+\cdot}$ and $H_2(2H^{\cdot})$. The shape of the IKES peak corresponding to this collision-induced reaction indicates that a large range of kinetic energies is released, the maximum being estimated as approximately 5 eV. The large energy loss, Q', indicates that the process corresponds to product formation in an excited state which is also the major process in unimolecular fragmentation as indicated from the shape of the ionization efficiency curve. The wide range of kinetic energies released (and correspondingly of internal energies retained) indicates that the accepted conclusion that this reaction occurs with generation of two hydrogen atoms may be incorrect. Rather a hydrogen molecule, which may be in a high vibrational state, is the product. The ion kinetic energy experiments agree with earlier conclusions that the product $O^{+\cdot}$ is excited, but it is not possible to decide between the ^2D and the ^2P states of this ion. Thus, taking the heat of formation of $H_2O^{+\cdot}$ as 233 kcal. mole^{-1}, the minimum energy acquired on collision as 18 eV and the maximum energy release corresponding to this excitation as 5 eV, one calculates $\Delta H_f(O^{+\cdot}) = 533$ kcal. mole^{-1}. This calculation is very approximate and is in best agreement with $\Delta H_f[O^{+\cdot}(^2P)] = 489$ kcal. mole^{-1}.

STEREOCHEMISTRY

This approach to ion structures is an outgrowth of the application of mass spectrometry to problems of molecular stereochemistry[65, 206]. It is an implicit assumption in most applications of this method of molecule structure determination that the structural integrity of the neutral molecule is retained in the fragmenting ion. When used as an ion structure probe, the steps involved are essentially as follows: two stereoisomers of known stereochemistry are examined and the relative importance of some elimination reaction, $M^{+\cdot} \rightarrow (M-HX)^{+\cdot}$, is determined. If the ease of elimination is substantially different in the two compounds and in the direction predicted from the minimum distance to which H and X can approach in the neutral molecule, then the tentative conclusion must be that the molecular ions have the same stereo structures as do the corresponding molecules.

Two general characteristics of the ion structure method may be noted before specific applications are discussed. First, the method is still crude since it is frequently not possible to do more than specify which of two isomers will undergo a given elimination reaction more readily — the actual ratio of ion abundances cannot be defined. For this reason it is possible that erroneous conclusions might be drawn — an unsuspected ring-opened ion structure might give the same abundance order as the cyclic ion and hence not be detected. Such errors can usually be avoided, particularly if specifically deuteriated compounds are used. A second point concerns the time scale employed in this type of experiment. It has been pointed out that there are advantages for distinguishing the mass spectra of stereoisomers in working under low-energy conditions, *i.e.* low source temperature and low electron energy. Specifically, isomerization and further fragmentation seem to be minimized. Most applications have not taken the next step and employed decompositions of metastable ions rather than those occurring in the source. This may actually prove fortunate since, with the extra time allowed, molecular ion isomerization could be enhanced if a suitable low-energy pathway were to exist.

Most of the studies on stereochemistry by mass spectrometry have focused on the elimination of water from alcohols. Many examples from the steroid field have been studied and axial hydroxyls invariably are found to be more easily lost than equatorial hydroxyls[283]. In some cases the differences in abundance are as much as two orders of magnitude[70]. One can conclude from this work and many similar studies that some fraction, at least, of the fragmenting ions has not undergone ring opening or epimerization. The central fact is that it is the selective orientation of two groups in space (*e.g.* H and OH in H_2O elimination) which determines the facility with which an elimination reaction proceeds; this has been stressed by Meyerson and Weitkamp[206]. More recently, particularly direct evidence for this hypothesis was found by Fenselau and Robinson[107] who studied steroidal diols and found that a prominent peak corresponding to elimination of water from molecular ions in the mass spectrum correlated with strong intramolecular hydrogen bonding as established by infrared spectrometry. This constitutes substantial evidence that the structure of the fragmenting ion is very similar to that of the neutral molecule. In similar fashion, Brown and coworkers[212] found that the elimination of H_2O and CH_3OH from 1,2-diphenylethanol and its methyl ether studied by mass spectrometry shows a stereoselectivity which is just that expected if bulky group (phenyl) interactions are to be avoided. Again, similarity between the structures of the ground-state molecule and the reacting ion is implied.

An interesting advance in methods of determining ion structures

has been made by Green *et al.*[122]. The essence of their idea is to
compare the behavior of a molecular ion and a species in solution as
regards the stereochemistry of their reactions. If in their dynamic
behavior the two species make the same stereochemical "choices",
then a structural relationship is evident. This can in fact be made a
stringent test by providing subtle differences in stereochemical choices.
In order to probe the structure of the 2-hexanol molecular ion,
each diastereotopic hydrogen at C-5 was in turn substituted by
deuterium. Now, it is well known that electron-induced dehydration
is highly specific (at least in the formation of detectable daughter
ions) and involves a C-5 hydrogen[29, 205]. The experiment by Green
et al. allowed the determination of which C-5 hydrogen was preferentially lost and allowed comparison of this result with that found
for an analogous solution reaction, the abstraction of a C-5 hydrogen
from the alkoxy radical generated from 2-hexanol under Barton reaction conditions. Both the mass spectrometric and the model solution
reaction showed preferential elimination of the same diastereotopic
hydrogen atom, indicating that the molecular ion must bear a considerable structural resemblance to the alkoxy radical. The similarities
between the two transition states are illustrated below.

Similarly, Green *et al.*[121] have shown that electron-impact and
pyrolytic elimination of acetic acid from alkyl acetates follow the
same conformational course and the conclusion again is appropriate
that the gaseous ion and the neutral ester must have identical structures differing only perhaps in bond lengths.

While there may be little scope for gross differences in structure
between ion and neutral in the acyclic chlorides, alcohols and acetates just discussed, in cyclic compounds ring opening may or may
not occur. The elimination of water from cyclohexanol and related
compounds has been an area of considerable mass spectrometric
effort, with the ring opening question central to the ion structure
problem. It has been shown[120] by stereospecific deuterium labeling
that both forms are present and both eliminate water. The cyclic ion
undergoes site and stereospecific 1,4-elimination of water while
1,3-elimination is preceded by ring opening, as indicated primarily by

the non-stereospecificity of this reaction.

In eliminating water, acyclic alcohols coil back on themselves. This coiling of ions is of interest as a means whereby charge might be stabilized in the gas phase. In bifunctional compounds, in particular, internal solvation of this type could be expected to provide considerable thermochemical stability. This topic has been discussed by Meyerson and Leitch[207] and while the concept is very attractive, the direct evidence is still limited, although the idea does help rationalize the fragmentation behavior of some long-chain bifunctional aliphatic compounds.

A particularly novel observation bearing on the ion structure problem concerns mass spectral evidence found for the formation of catenanes from large cyclic polyenes[25, 278]. An intramolecular metathesis reaction can occur to give either the cyclic olefin or the catenane depending on whether the initial olefin is twisted through 180° or not.

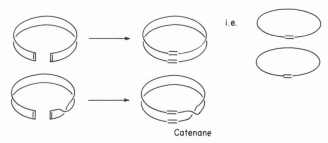

Catenane

The presence of the catenane was detected by the sudden reappearance of abundant ions corresponding to a cyclic mono olefin as the source temperature was raised. (At low temperature, molecular ions of this olefin were abundant but at higher temperatures, they were largely absent.) Apparently, at sufficiently high temperatures, ring opening of the catenane ion occurs. Thus, it seems that cyclic olefins give cyclic molecular ions which, given sufficient internal energy, can ring open.

Molecular ions of other classes of compounds, especially highly unsaturated and aromatic compounds, present more of a structural problem than do monofunctional aliphatics and alicyclics, and the stereochemical approach has made almost no inroads here. In some conjugated dienes, however, evidence has been presented for retention of the molecular structure in the molecular ion. In a comparison of the relative ease of loss of methanol from the isomeric esters

$$C_2H_5 \quad C_2H_5$$

$$H_3CO_2C \overset{\diagup}{\diagdown} CO_2CH_3$$

$$C_2H_5 \quad C_2H_5$$

$$H_3CO_2C \quad CO_2CH_3$$

the *trans, trans* isomer showed significant methanol elimination while the *cis, cis* compound did not[118]. Only in the former case does the geometry of the neutral molecule allow close approach of the methoxyl and allylic hydrogen moieties. This result is particularly significant since mere rotation about a double bond is required for the *cis, cis* isomer to lose methanol. The fact that the elimination does not occur is testimony both to the similarity of ionic and neutral molecule structure and to the absence of isomerization even, in this case, in molecular ions formed using 70 eV electrons.

ION CYCLOTRON RESONANCE AND ION—MOLECULE REACTIONS

The use of ion cyclotron resonance spectrometry in determining the structures of gaseous ions has several unique features. The method depends on assigning ion structure on the basis of reactivity with neutral molecules. One is therefore studying stable rather than reactive ions. The ions which are sampled will generally have been formed by electron impact, but their internal energies will be lower than ions studied in conventional mass spectrometers because of the longer average ion lifetime in the ICR instrument before reaction. Another consequence of this delay is that any isomerization is likely to be complete and the thermodynamically favored isomer is likely to predominate (see discussion of $C_3H_7^+$ below).

Other important characteristics of ICR experiments[13, 14] are that low kinetic energy ion—molecule reactions are studied, double resonance techniques allow complex ion—molecule reaction sequences to be resolved and the exo- or endothermicity of many reactions may be determined[19] and, finally, the use of isotopic species yields information on the intermediates in these ionic reactions.

The characterization of an unknown ion using several neutral reagents can be illustrated in the case of the $C_2H_5O^+$ ion. Consider three sources of this ion: (*i*) protonation of ethylene oxide, (*ii*) protonation of acetaldehyde and (*iii*) fragmentation (—H·) of the dimethyl ether molecular ion. Experimentally, it is observed that the ions formed by (*i*) and (*ii*) transfer H^+ to NH_3 and to isobutylene but not to propene, while that generated by method (*iii*) transfers only to NH_3 (ref. 20). The ion generated from dimethyl ether can also be distinguished from the others by hydride abstraction and methyl cation transfer reactions. On the basis of these results, it has been suggested that two structural forms are involved in these reactions, the ion a being formed from dimethyl ether and b being formed from the other reactants.

$$CH_2 = \overset{+}{O} - CH_3 \qquad\qquad H_3C - C\overset{\displaystyle \overset{+}{O}H}{\underset{H}{\bigg\backslash}}$$

     a   b

The protonated ethylene oxide is believed to isomerize to the more stable protonated acetaldehyde structure prior to or during reaction. It should be emphasized that the problem of distinguishing differences in structure from differences in internal energy arises here, just as it does in interpreting the results of unimolecular ionic fragmentations. Substantial variation in the reactivity of ion b with internal energy was observed by Beauchamp and Dunbar[20]. The lower internal energies of ions studied by ICR must be balanced against the fact that, given the intimate nature of the ion—molecule collision complex, even small differences in ion internal energies may have pronounced effects on overcoming activation barriers. Generalizations regarding the importance of internal energy must await the acquisition of further data but it is, potentially, a serious limitation in the use of ICR in characterizing ion structures.

The approach to ion structure just illustrated for $C_2H_5O^+$ has been used in several other cases and just a few examples will suffice. $C_3H_6^{+\cdot}$ ions can be divided into two classes[123] depending on whether or not they undergo the following two reactions with ammonia.

$$C_3H_6^{+\cdot} + NH_3 \rightarrow CH_4N^+ + C_2H_5\cdot$$

$$C_3H_6^{+\cdot} + NH_3 \rightarrow CH_5N^{+\cdot} + C_2H_4$$

Because of the all-or-nothing nature of this case the classification almost certainly reflects structural rather than energetic differences.

$C_3H_7^+$ ions are of considerable interest theoretically and have been studied experimentally by radiolysis and in the course of solution carbonium ion studies, as well as by mass spectrometry. McLafferty and coworkers[180] make use of partially deuteriated $C_3H_7^+$ and follow H^+ and D^+ tranfer to such bases as methanol, propanol and acetone. By measuring the isotope effects operating in their system, they were able to assess the degree of isomerization prior to proton transfer. n-Propyl cations isomerized almost completely, and i-propyl cations hardly at all, indicating that the process $n\text{-}Pr^+ \to i\text{-}Pr^+$ is thermodynamically favored. This approach could not be applied to ring-protonated cyclopropane (ring-$C_3H_7^+$) since hydrogen scrambling is expected to be rapid within this ion. It was shown, however, that ring-$C_3H_7^+$ had an identical proton affinity to i-$C_3H_7^+$, strongly suggesting prior isomerization to the i-$C_3H_7^+$ structure.

A problem of ion structure which has been solved by ion cyclotron resonance concerns the ions formed by the McLafferty rearrangement in ketones, particularly the ion formed by two successive alkene eliminations from a dialkyl ketone[100, 104]. Ionization potential data were interpreted[203] as implying the enol structure, c, for the product formed by loss of two olefinic neutral fragments, but early studies on metastable ion abundances favored the oxonium structure, d.

c   d

These results are not necessarily incompatible since they concern threshold and reactive ions, respectively. However, later metastable ion data[191] supported the enol structure for the product.

The ICR method depends upon finding an ion—molecule reaction given by either the keto ion or the enol but not by both. Reaction with 2-hexanone is such a process. The keto ion formed by direct ionization of acetone reacts as shown with the 2-hexanone reagent.

The isomeric enol ion formed from 1-methylcyclobutanol does not react. Neither the product of McLafferty rearrangement nor the product resulting from the second alkene elimination of dialkylketones reacts with 2-hexanone under comparable conditions, indicating that neither ion possesses the keto structure.

The ICR methods so far discussed have probed the structures of unknown ions by considering the reaction, or lack of it, with given neutral molecules. As a variation upon this theme, the reactant ion and neutral might both be of known structure while that of the ion—molecule product might be in question. Consider for example, the gas-phase acetylation of phenol[27]. The product might possess either of the structures e or f.

e   f

The actual structure was shown to be e by considering the ease of acetylation of a series of o,o-dialkylphenols. Acetylation decreased down the series R = H, CH_3, C_2H_5 and ceased entirely for R = i-C_3H_7, t-C_4H_9. m,m-Di-t-butylphenol was readily acetylated proving that a steric not an electronic effect was involved.

Our final topic in this treatment of ICR concerns the determination of transition-state structures in ion—molecule reactions. As comparisons between gas-phase ion—molecule reactions and solution chemistry have become more common, this type of development has become significant. The S_N2 reaction $Cl^- + RBr \rightarrow RCl + Br^-$ has been studied in the gas phase and the rate constants determined[128]. The relative order of reactivity is very different from that in solution and even groups such as adamantyl and t-butyl in which rear attack is prevented, react with ease. The nucleophile must therefore attack from the front in the gas phase, the transition state being of the type g rather than h.

g   h

Another example[26] of ion and transition-state characterization by ICR concerns reactions which are analogous to electrophilic aromatic substitution as observed in solution (*i.e.* $E^+ + ArH \rightarrow ArHE^+ \rightarrow ArE + H^+$). The difference is that in the gas phase, either the first step alone occurs, as for example in the nitration

$$C_6D_6 + H_2ONO_2^+ \rightarrow C_6D_6NO_2^+ + H_2O$$

or it is followed by an alternative second step as in the chloromethylation

Parenthetically, it may be noted that the nitration reaction with methyl nitrite is undergone by the toluene molecular ion but not by the isomeric cycloheptatriene and norbornadiene molecular ions. This has been interpreted as indicating that the non-decomposing toluene molecular ions do not isomerize or ring expand[142].

A SPECIFIC EXAMPLE OF AN ION STRUCTURE DETERMINATION

In this section we consider a specific example of an ion structure determination which has been studied using several of the techniques discussed above.

The structure of the benzene molecular ion provides an example of a problem which has been studied by a variety of techniques. As in other questions of this type, the structures of stable and reactive ions must be considered separately and only the reactive $C_6H_6^{+\cdot}$ ion generated from benzene will be considered here. The comparison by Momigny *et al.*[208] of the mass spectrum of benzene and those of some of its open-chain isomers (the hexadienynes) provided qualitative evidence that the reactive form of the benzene molecular ion had an acyclic structure. Andlauer and Ottinger's experimental k *vs.* ϵ data[4, 5] for the major reactions of the benzene molecular ion indicate that the ion which loses C_2H_2 (and $C_3H_3^{\cdot}$) has a different structure from that which loses $H^{\cdot}$ (and H_2). This conclusion is based on the k *vs.* ϵ curves which show that the fragmentations cannot be competitive. The result suggests that there exists both a linear and a cyclic reactive form of $C_6H_6^{+\cdot}$. The postulation of two structures for the reactive $C_6H_6^{+\cdot}$ ions is hard to reconcile with photoionization efficiency results which show that, within experimental error, the reactions leading to $(M-H)^+$, $(M-H_2)^{+\cdot}$, $(M-C_2H_2)^{+\cdot}$ and $(M-C_3H_3)^+$ have identical appearance potentials[83]. The idea that there is only a single reactive structure for $C_6H_6^{+\cdot}$ also emerges from the study of the behavior of $C_6H_6^{+\cdot}$ ions generated as fragment ions from tropone, anisole, styrene, cyclooctatetraene and m-dimethoxy-benzene. The ions so generated all behave similarly in undergoing loss of $H^\cdot$ and loss of C_2H_2.

Since doubly charged ions can be efficiently converted to singly charged ions by charge exchange in the presence of a collision gas, and since evidence has already been provided (see p.191) to show that the reactive form of the doubly charged molecular ion of benzene is acyclic, the fragmentation of $C_6H_6^{+\cdot}$ generated from $C_6H_6^{2+}$ was monitored[159]. It must be noted that most of the $C_6H_6^{2+}$ ions which undergo charge exchange in the field-free region to give $C_6H_6^{+\cdot}$ will be non-reactive and it cannot be shown directly that these ions are acyclic. However, it may be noted that the cyclic structure would represent a high-energy anti-aromatic four π-electron system which, if formed in a vertical ionization process, would be expected to undergo ring opening readily.

The linear $C_6H_6^{+\cdot}$ ion formed as described underwent both C_2H_2 loss and $H^\cdot$ loss, and the energy release accompanying these reactions was similar to that measured for the benzene molecular ion. These results are inconsistent with the view that there are two isolated states of the reacting benzene molecular ion and they indicate that the reacting benzene molecular ion has an acyclic structure.

Theory of mass spectra

"A general flavour of mild decay
But nothing local, as one might say"

O.W. Holmes, *The Deacon's Masterpiece*

INTRODUCTION

The statistical approach to the theory of mass spectra owed its origin primarily to two main considerations which need to be recognized at the outset. The first of these depends on the empirical evidence that polyatomic positive ions appear not to dissociate immediately they are formed by electron impact, thus permitting the internal energy to be distributed amongst many internal degrees of freedom. The second consideration is essentially one of an inherent deficiency of detailed understanding; idealistically, as the mathematician Laplace conjectured, given sufficiently detailed knowledge of the instantaneous state of the universe its whole subsequent course should be predictable. But in practice we do not possess such detailed knowledge even in the case of a system no larger than a polyatomic ion, owing to the large number of excited electronic states implicated and concerning which little at all is known for certain. Wallenstein and coworkers[265] were the first to draw attention to these circumstances concerning the theory of mass spectra. It will not be the purpose of the appendix to develop *ab initio* the statistical theory from its basic premises; this has already been done by several authors. It will suffice for our purpose to touch upon those features of the theory which appear to us to be of especial importance for understanding the properties of metastable ions, and without which much of the discussion in the various chapters of this book may appear incomplete. For facility of exposition, we shall also refer to the RRKM version [158, 176, 228] of the statistical theory which, though formally quite different from the quasi-equilibrium theory (QET) [231], contains all its essential physical features in a more readily assimilable form.

It is usual to start these discussions by reminding the reader that an impact electron passing in close proximity (*i.e.* within molecular scale of distances) to a particular molecule must have completed its interaction within a time as short as 10^{-16} sec. As discussed in Chapter 1, this is the order of time taken for a 50 eV electron to travel several Ångstroms. In this very short time the heavy nuclei cannot have moved perceptibly and thus the electron excitations may be considered to be ·vertical processes satisfying the Franck—Condon principle. This leads to the important conclusion that the nuclei are originally not in their most favorable locations minimizing the energy of the ion, but that they are substantially in the configuration minimizing the energy of the molecule before ionization. Consequently, they will commence to execute oscillation in an attempt to restore equilibrium. Provided that electronic states differing little in energy can exist, the vibrational energy should be sufficient to produce radiationless energy transfer between such closely situated potential energy surfaces. The magnitude of the probable mean energy spacing is considered in the later mathematical treatment. Eventually, much of the electronic energy deposited by the electron-impact process will, in view of this mechanism of energy transfer, be converted into vibrational energy distributed amongst the numerous internal degrees of freedom, the ultimate electronic state being, of course, that of least energy (but see later).

The essential task of the QET is to evaluate the relative probabilities of particular assignments of excess internal energy amongst the available degrees of freedom. This assumes, as required at the outset, that a typical ion does not dissociate instantaneously but can exist for a time sufficient for the requisite internal energy transfer processes to be effected. Empirical evidence for this assumption is provided by such well-established properties as the absence of any direct correlation between the number of bonds of particular types and the relative abundance of corresponding fragment ions. To take the simplest example, the loss of a single hydrogen from paraffin hydrocarbons shows no simple direct dependence on the number of CH bonds. If dissociation followed immediately the ejection of an electron from a CH bond, consideration of the equivalence of many such bonds would imply equivalence of ionization probability and a direct dependence of the overall result on the number of such bonds. Further evidence follows from such results as the close correspondence of the total ionization for different isomeric forms, to be contrasted with the often striking differences in fragmentation behavior.

These and other features of the QET will be considered in the course of this treatment. Some mathematical aspects concerning the statistical distribution of electronic states, the RRKM exposition of internal energy partitioning and the state density function approach will be discussed separately later. The remainder of the appendix will consist of a variety of topics encountered in applications of the theory, with particular relevance to metastable ions. Amongst those discussed are the active site concept applied to triggering mechanisms, Fermi resonance, the significance of particular types of excitation energy, specific unimolecular rate processes *vis-à-vis* those where a temperature is defined, the significance of the so-called kinetic shift and equipartition effects in homologous series. Later sections will deal with simplified versions of QET specifically designed by Williams and co-workers for application to complex organic compounds, isotope effects, rates of crossing of energy surfaces and correlation rules for the fragmentation of highly symmetrical positive ions.

THE STATISTICAL HYPOTHESIS

The essential requirement of the quasi-equilibrium or statistical theory of mass spectra is that dissociation of a molecular or fragment ion is a relatively slow process compared with the rate of conversion of energy between the various degrees of freedom, electronic, vibrational and rotational. Taken together with the fundamental principle involved in statistical thermodynamics, namely that all degrees of freedom in equilibrium may be treated on an equal footing of equal *a priori* probability, this implies that the relative probabilities of particular modes of decomposition may be determined simply by counting up the numbers of distinguishable allocations of energy. The numbers are not, however, unrestricted since there is a constraint imposed by the requirements that the total available energy is fixed at the instant of electron impact by the conditions of such impact. Loosely speaking, if the impact electron strikes the center of a bond, it will excite much more internal energy than if it passes by on one side of the bond at a distance of several atomic radii. The impact is due to an electron wave-packet which is not a localized charge at a geometrical point but is spread out in space and time and thus it is not strictly correct to refer to a particular point of impact. But as Born emphasized, the electron wave can be regarded as measuring the probability of observing the point electron at some point of space, or equivalently — in Schrödinger's view — the wave itself represents a single electron that is delocalized or smeared out as an electron cloud.

It makes no difference to the final result, so it is convenient to accept the first viewpoint and regard the impact as due to point charges emerging from the hot filament under thermionic emission. There will thus be a distribution of excited molecules, some remaining as neutral species, others consisting of positive ions having various amounts of excitation energy. When applying statistical criteria to any particular group of ions, there is imposed an internal energy constraint appropriate to that group. But it is important to recognize that all one is able to do with the energy in a particular group is to distribute it internally in many different ways. Its total amount will only be different for other groups of ions that underwent a different kind of excitation. This means that the statistical problem is divided into two main parts, the first involving the specific rate for some given total internal energy and the second which considers the excitation function or profile of internal energies. The specific rates may not be averaged over this energy distribution unless it be sufficient to evaluate the total decomposition over all places and all times. In the mass spectrometer one is usually particularly interested in dissociations that occur within a specific time range (typically 5 μsec after electron impact for a metastable ion) or within a given limit of time (those that dissociate, for example, before entering the accelerating field). This means that the unimolecular decomposition for a specific rate must be evaluated as e^{-kt} or $(e^{-kt_1} - e^{-kt_2})$ and then averaged over these expressions rather than over the rate constants themselves. Only for very small values of kt where $e^{-kt_1} - e^{-kt_2} \cong k(t_2 - t_1)$ would the averaging processes become equivalent.

There is a further subtlety about energy distribution processes which comply with the equilibrium principle or that of equal *a priori* probabilities. To achieve this equilibrium the various potential energy surfaces, corresponding to differing amounts of electronic excitation energy, must be sufficiently close together in the configuration space of the molecule to permit crossing from one surface to another, otherwise the only way of equilibrating the energy would be by the comparatively slow (10^{-8} sec or longer) process of radiational transfer. If radiationless transitions by surface crossings are to be freely available, there is nothing in principle to preclude a traffic both ways involving the reoccupation of at least a few of the lower-lying energy surfaces. That is to say, it is not simply a matter of the energy cascading down to the very lowest-lying energy surface unless such a surface lies much lower in energy than all the others and in that case the mechanism for transfer in the available time does not exist. Nevertheless, the statistical nature of the phenomenon requires that the state density functions will allow for the occupation of all states, involving electronic excitations as well as other forms. In practice, it is commonly assumed that the lowest potential energy surface plays a major role since the occupation of the different electronic states will be in the ratio of their densities of states which greatly favors this state. Hence, despite the above complications, it is conventional to regard the important features of the mass spectra as being dependent on motion over a single (usually the lowest) potential energy surface. But Kropf *et al.* [164] in their discussion of propane and dideuteriopropane point out that there is a discrepancy unless the transition state frequencies for certain primary CH and CD dissociations are reduced, an effect that they attribute to the next higher potential surface. It is therefore clear that there is nothing inevitable about the assumption of dissociation over the lowest-lying potential energy surface; it is not a necessary feature of the statistical theory as such.

This may be a convenient point to refer to the active site concept and the triggering mechanism attributed to the presence of ion-radical sites [186]. This really amounts to the choice of a certain valence bond structure in which the unpaired electron is visualized as localized, typically on a heteroatom, such as the oxygen atom of a carbonyl group. Thus, transfer of a γ-hydrogen in aliphatic ketones is visualized as triggered by the oxygen cation.

If there is only a single way of writing a structural formula of the ground electronic state then it may be supposed that this particular electron pairing presents a good approximation to the ground state of the radical ion, and the only degeneracy permissible is that to be attributed to the unpaired spin, which can be either α or β spin. This means that there is a single potential energy surface or more exactly two energy surfaces with an infinitesimal separation. The single structure hypothesis provides no other low-lying potential surface and thus precludes any provision for the manifold intersections necessary for the statistical theory to be valid. The existence of two or more valence bond structures implies an equal number of eigenstates, for the dimensionality of the interaction matrix, which is itself equal to the number of valence bond structures, determines the number of closely positioned potential energy surfaces. Whilst the single structure associated with an active site is able to provide one of the important features of the statistical approach, namely that it provides a higher volume in phase space than any alternative structure because of its favorably low electronic energy, yet it makes it difficult to satisfy the other requirement since radiationless energy transfer between widely spaced electronic states may be impossible. Instead of competitive reactions, it is now possible to have simultaneous reactions over a number of distinct and non-interacting potential energy surfaces. It may be significant that charge exchange experiments [256, 260] have recently provided evidence for fragmentation from isolated electronic states in dialkyl ketones. This suggests that the ground electronic state may primarily be implicated in the fragmentation of ketones but that when the much higher electronic states are populated in the ionization process, they subsequently undergo an independent ion chemistry.

The conclusion that a metastable ion must (for comparable frequency factors) choose the route to dissociations of lowest activation energy need not apply in the case of non-interacting surfaces if there is no path enabling dissociations of higher activation energy to implicate a lower-lying potential energy surface. What must now be considered is a pair of simultaneous reactions proceeding as parallel processes. If the rate constants are k_1 and k_2 for the non-interacting potential energy surfaces and the initial precursor ion abundances referring to these surfaces are a_1 and a_2, then in the time interval τ to $(\tau+d\tau)$ the numbers of fragmentations are, respectively, $a_1(e^{-k_1\tau}-e^{-k_1(\tau+d\tau)})$ and $a_2(e^{-k_2\tau}-e^{-k_2(\tau+d\tau)})$ which approximately equal $a_1 k_1 d\tau e^{-k_1\tau}$ and $a_2 k_2 d\tau e^{-k_2\tau}$. Let us consider a metastable transition such that $\tau \cong 10^{-5}$ sec. Then if k_1 and k_2 are both of the order of 10^5 sec^{-1}, both dissociation processes will lead to significant amounts of fragmentation within this time interval. If the activation energy for the lower energy surface is reduced with a corresponding increase in k_1 to something much larger than 10^5 sec^{-1}, the dissociation, $a_1 k_1 d\tau e^{-k_1\tau}$ in the prescribed interval is reduced to insignificance. Thus if k_1 becomes ten times larger, this fragmentation is attenuated by the factor $10e^{-10}$. But the parallel process is in no way affected and the *higher* activation energy process is responsible for the metastable ions, contrary to the conventional conclusions of the QET. When crossings between the two surfaces are readily available, it is as though there were a single effective energy surface and the dissociations taking place by the two mechanisms yield respectively $a_1 k_1 d\tau e^{-(k_1+k_2)\tau}$ and $a_1 k_2 d\tau e^{-(k_1+k_2)\tau}$.

In this case $(k_1 + k_2)$ has to be close to 10^5 sec^{-1} otherwise metastables will not be observed at all. And it is essential that there is no competitive route of lower activation energy which would have the capability of raising $\Sigma_i k_i$ to appreciably more than 10^5 sec^{-1}. Thus in the QET it is always the route to a particular dissociation of highest rate and generally lower activation energy that determines the metastable ions.

It is possible to relax the equilibrium requirements by stipulating that only partial equilibrium is attained. Whilst there may be several potential energy surfaces, each one may then be considered to achieve its own state of equilibrium independently of the others [258]. This leads to changes in some conclusions such as the one mentioned above. A further difficulty is that the requisite mechanism for efficient internal energy transfer is no longer available in the absence of extensive potential surface intersections. A way round this difficulty is to invoke Fermi (and perhaps Coriolis) resonance [233]. Briefly, this means that for polyatomic ions, accidental degeneracy can occur for vibrational states of the same or different species, but for highly symmetrical ions, only those states of the same species can perturb each other as a result of anharmonicity of the vibrational motions. For diatomic molecules, Fermi resonance is only possible for vibrational levels belonging to different electronic states [133]. Fortunately, for polyatomic systems, the resonance can occur for the same electronic state and thus a localized bond vibration can distribute its energy amongst other suitable normal modes if anharmonicity is sufficiently extensive. It is in the nature of metastable ions that only marginally enough energy is available to lead to fragmentation. Thus, the Fermi resonance must be an inefficient energy transfer mechanism because a sufficient number of quanta are not available to produce anharmonic vibrations simultaneously in several oscillators. The considerable changes in geometry required to produce many transition states seem to demand something more drastic than slight perturbations resulting from anharmonicity. In effect this mechanism allows only secondary changes in geometry while leaving the skeletal framework somewhat distorted but still recognizably the same structure as that of the precursor ion. For gross distortion leading to rearrangement processes, it becomes necessary to contemplate ion structures that are significantly dissimilar to the ground-state forms and, if an equilibrium hypothesis is to retain any real meaning, it must include the numerous crossings of potential energy surfaces contemplated in the originally formulated QET hypothesis.

As an additional complication to the problem of deciding which particular mechanism of energy equilibration is primarily involved, there remains that of assessing the validity of mathematical approximations involved in arriving at simplified rate expressions. This is a separate problem which is discussed later on in this appendix where it is shown that refinements must be introduced in counting the number of configurations where there is only marginally sufficient energy for fragmentation. It has been remarked earlier in this appendix that recourse to a statistical treatment for polyatomic ions is ineluctable in practice. A statistical treatment is only permissible providing that an effective mechanism leading to a state of equilibrium is available, though it is not always easy to see *a priori* which particular mechanism is predominant in any particular situation. In the absence of any such mechanism the conclusions will have little more than a formal validity loosely described by an "effective number" of oscillators, not having quite the same truly physical connotation as a number of "effective oscillators". Whilst it may be rather oversimplifying the situation, it would appear from general considerations of average spacing between potential energy surfaces, to be considered in the mathematical section, that for moderately large polyatomic ions, the necessary extent of surface intersection is available even in the threshold region. But for small ions, such as often occur in secondary fragmentation processes, Fermi resonance and other quite subtle mechanisms acquire a greater importance.

It is important to recognize certain features of the various types of excitation energy. It is usual to speak of electronic energy, vibrational energy and energy of structural rearrangement as though these were distinct types of excitation, though this is a somewhat arbitrary distinction which ceases to be meaningful for extreme states of internal energy. Thus the whole basis of separating electronic and vibrational factors in the wave functions depends on the adiabatic (otherwise the Born—Oppenheimer) principle [68]. This is generally valid because of the great disparity in mass of electron and nuclei and in view of the excitation of a few vibrational quanta only so that the nuclei execute their simple harmonic oscillations near the bottom of a parabolic potential well. Their mean position is virtually the same as for the zero-point level, appropriate to the case of zero internal excitation. But the extreme states of internal excitation contemplated in Fermi resonance imply a potential well that is significantly asymmetrical and the mean positions of the nuclei depart significantly from their ground-state configuration. In such cases there may be appreciable interaction between vibrational and electronic motion so that the separation is no longer valid. It is a characteristic feature of different electronic states that they determine somewhat different nuclear configurations. Anharmonicity achieves the same results by vibrational excitation comparable in effect with that of change of electronic state and we must therefore be prepared for some weakening of the adiabatic principle itself because of a state of comparability of energies of two different forms of excitation being reached. Continuing along the same lines of argument, one is led to the conclusion that the grosser distortions and rearrangements of precursor ions in many metastable ion fragmentations, sometimes sufficient to merit the description of isomeric change, are substantially another case of electronic excitation where interaction of nuclei has also simultaneously undergone change. Such changes are not normally described as conventional excitation of higher electronic states because this has the connotation of an adiabatic principle where the nuclei retain the same or only slightly disturbed locations. But the logical distinction becomes somewhat dubious in cases where the rearrangements are accessible during the time of an observation. The relevant deformations then become feasible operations[1 7 2] and the point group of the ion should be augmented to include additional operations, after the manner of a Schrödinger supergroup[3]. It would then become logically consistent to refer to electronic excited states based on instantaneous configurations representing a wider range of nuclear transformations than those contemplated in the ordinary point groups and leading to various extended normal modes of vibration. In this wider sense, the concept of excited state may be extended to the isomeric changes occurring in many ion fragmentation reactions. In the ultimate analysis, all that can be said is that certain configurations of electrons and atomic nuclei lead to stationary total energy and that the excitation energy depends on both in a complex configuration space for which the ordinary concept of a potential surface with potential wells around particular nuclei is at best a rather naive one. Thus we must regard as eigenstates generalized functions of electron and nuclear co-ordinates that imply sharp energy values as determined by the Hamiltonian operator.

RATE CONSTANT DEPENDENCE UPON INTERNAL ENERGY

The value of the rate constant, k, is considered in the mathematical section. It has often been expressed in simplified form as

$$k = \nu \left(\frac{\epsilon - \epsilon_0}{\epsilon} \right)^{s-1}$$

where ϵ is the internal energy of the ion undergoing fragmentation, s is the number of effective oscillators in the ion, ϵ_0 is the activation energy and ν the frequency factor [233]. This rate expression, to which we have recourse extensively in most semi-quantitative treatments of unimolecular fragmentation processes, is derived in the mathematical section both as a simplified version of the QET rate expression and on the basis of the RRKM statistical model.

It is clear that this shows a quasi-exponential dependence on the activation energy, for

$$\left(\frac{\epsilon-\epsilon_0}{\epsilon}\right)^{s-1} = \exp\left[(s-1)\ln\left(1-\frac{\epsilon_0}{\epsilon}\right)\right]$$

which for $\epsilon_0/\epsilon \ll 1$ may be expressed as $\exp(-(s-1)\epsilon_0/\epsilon)$. Comparing this with the form of the expression for the Arrhenius rate constant, we notice that $\epsilon = (s-1)RT$ where T is the temperature for thermal equilibrium in the latter case. Incidentally, it may be appreciated that the relative weights of two closely situated electronic energy levels is approximately $\exp\{-\Delta E/(RT)\}=\exp(-(s-1)\Delta\epsilon/\epsilon)$.Thus in a typical case of 50 oscillators and 10 eV excitation energy, states separated by 0.2 eV must occur to a significant extent in the final equilibrium. For metastable ions, the energy ϵ is only marginally greater than ϵ_0 so that the exponential form is a relatively unsatisfactory approximation. However, there is a compensating effect in that the number of effective oscillators is lower than the actual number when $\epsilon_0/\epsilon\cong1$ since the approximate method for counting the density of states leads to a gross overestimate. The simple form of k can only be retained near the threshold energy for fragmentation, provided s is itself made an increasing function of ϵ. Thus, we ought to write

$$(s-1)\ln(1-\epsilon_0/\epsilon) = \frac{-(s-1)\epsilon_0}{\epsilon}(1+\tfrac{1}{2}\epsilon_0/\epsilon+\ldots.)$$

where s increases with increasing ϵ and thus, to some extent, compensates for the behavior of the bracketed expression which is itself a decreasing function of ϵ. Thus the Arrhenius form, where T may be introduced purely as a defined quantity, is actually a somewhat better representation of the rate constant than might be expected. But the exponential form does suggest that a useful representation of the energy dependence is in the logarithmic form

$$\ln k = \ln \nu - \frac{(s-1)\epsilon_0}{\epsilon}(1+\tfrac{1}{2}\epsilon_0/\epsilon+\ldots)$$

and this simulates closely a hyperbolic dependence on energy, with an asymptotic value

$$\ln k = \ln \nu$$

and a threshold value of $-\infty$.

The actual curves are often shown cut off when they reach the energy axis as in Fig. 96 so that the asymptotic approach to $-\infty$ such as is shown in Fig. 95 is excluded. In effect this only excludes the almost vertical part of the curve between $k = 0$ (the actual threshold value) and $k = 1$ (the nominal threshold value), for which the energy increment is given by $(\epsilon_0/10^{\mu/(s-1)})$ where $\nu = 10^\mu$.

We shall show in a later discussion of state density functions that the simplified rate expression ceases to have any real validity at threshold energies unless the number of oscillators is arbitrarily reduced. It might also be mentioned that the detailed theory leads, not to a zero threshold rate, but to a finite one as Vestal[262] has emphasized.

Writing k, for $J = 0$, in the form

$$k(\epsilon) = \frac{1}{h} \frac{W^{\ddagger}(\epsilon-\epsilon_0)}{\rho(\epsilon)}$$

where $W^{\ddagger}(\epsilon-\epsilon_0)$ is the number of states of the activated complex configuration with energy $\leqslant (\epsilon-\epsilon_0)$ and where $\rho(\epsilon)\, d\epsilon$ is the number of reactant ion states with energy in the range ϵ to $(\epsilon + d\epsilon)$, the limiting form is seen to be

$$k_{lim} = \frac{1}{h\rho(\epsilon_0)}$$

since the threshold value for $W^{\ddagger}(\epsilon-\epsilon_0)$ is not zero but unity. Thus, in an accurate sense, both assumptions about the minimum value of $k (i.e.\ 0$ or $1)$ are unrealistic. There is a finite limiting value at threshold energy and below this k must tend rapidly to a vanishingly small value $(\log k \to -\infty)$ as shown in Fig. 95. This is clearly the case, for if there is less internal energy present in the ion than the amount required to provide the activation energy, only a leakage through the barrier by a tunneling mechanism remains as a possibility. It can be shown that the rate then falls away exponentially (or nearly so) as a function of the energy deficiency and on a logarithmic scale this implies linear dependence. The almost vertical part of the logarithmic rate curve shown in the figure refers to this feature of the threshold energy region.

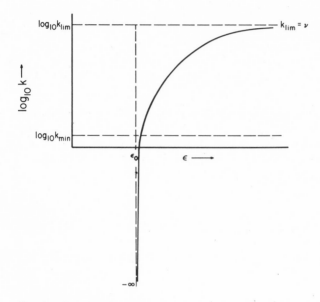

Fig. 95. Relationship between the rate constant for unimolecular fragmentation k, and the internal energy, ϵ, of a positively charged ion. The curve shows the threshold energy (activation energy ϵ_0) for reaction and the upper limit for the reaction rate constant.

Certain useful conclusions may be drawn from inspection of these logarithmic rate curves. If ϵ_0 and s are fixed in value but different frequency factors are considered, then two rate curves will be obtained, related to each other by a constant vertical displacement as shown in Fig. 96. If they were straight lines they would consequently be parallel, but they are not quite parallel in view of the curvature. Nevertheless, the curve of higher frequency factor lies everywhere above any similar curve of lower frequency factor. The energy range corresponding to metastable ions is greater for the lower frequency factor because the gradient is reduced in the appropriate region.

If s and ν are held constant but activation energies differ, the log k *versus* ϵ curves are again identical in shape, being related by a constant displacement along the energy axis equal to the activation energy difference as shown in Fig. 97.

Considering the effect of changing the frequency factor for a reaction as shown in Fig. 96, it can be seen that as the frequency factor is reduced (moving from curve (b) to curve (a)) the range of ϵ which corresponds to $\log_{10} k$ values lying between 4.5 and 5.5 increases. Thus, the relative abundance of the metastable ions increases. Considering the effect of changing the activation energy for a reaction, as

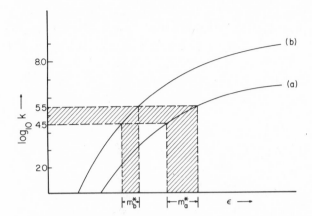

Fig. 96. Effect of different frequency factors on the range of internal energies over which the products of fragmentation of metastable ions can be observed. Curve (a) which has the lower frequency factor leads to the greater abundance of metastable ions (m_a^*). Note that the two curves are related by a constant vertical shift.

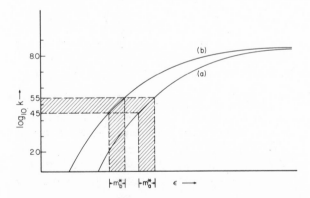

Fig. 97. Effect of different activation energies on the range of internal energies over which the products of fragmentation of metastable ions can be seen. Note that the two curves are related by a constant horizontal shift.

shown in Fig. 97, it can be seen that the relevant segment of the $\log_{10} k$ axis always intersects the curve in the same place and there is thus no change in the relative abundance of the metastable ions. (It should be emphasized that the two curves shown in Figs. 96 and 97 do not represent competitive reactions.) The subject is treated more precisely by Rabinovitch and Setser [225] and for a more detailed study reference should be made to their review.

The right-hand side of Fig. 98 shows the approximate regions of the rate curve over which stable ions, metastable ions and fragment ions occur [249]. These regions are not sharply defined; unimolecular decay implies that the average and individual lifetimes of particular ion species are different. The left-hand side of Fig. 98 illustrates the consequent appearance of more than one kind of ion at a particular value of the rate constant. The actual abundances of stable,

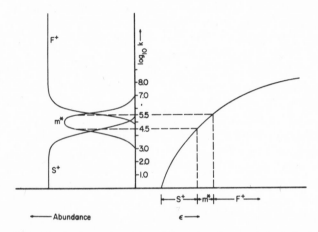

Fig. 98. Interrelationship of ion abundance, internal energy and reaction rate constant. The abundances of F^+, m* and S^+ give the relative probabilities that an ion will decompose in the source or in the field-free region or that it will reach the collector without fragmentation.

metastable and unstable ions will depend upon both the shape of the rate curve $k(\epsilon)$ and that of the energy distribution function $\rho(\epsilon)$. The abundance curves shown relate to a uniform distribution of internal energy.

The relative kinetic shift (compare p. 91) which measures the amount by which the total internal energy exceeds the energy of activation is given by

$$\frac{(\epsilon-\epsilon_0)}{\epsilon} = \left(\frac{k}{\nu}\right)^{1/(s-1)}$$

The absolute value of the kinetic shift $(\epsilon-\epsilon_0)$, (or $\epsilon^{\ddagger}$) is the extent to which the excitation energy in the transition state exceeds the activation energy determined by the height of the potential energy barrier in the fragmentation co-ordinate. Clearly for fixed s and k, this indicates that the relative kinetic shift is larger the lower the frequency factor. From the correlation between relative abundance and frequency factor, it follows that the kinetic shift and the relative abundance should increase or decrease together. The dependence may be slight because ν will only vary by a factor of the order of 10^4 from rearrangements on the one hand to simple cleavage reactions on the other. Since $(s-1)$ may be as large as, for example, 50, in typical polyatomic ions, the actual likely variation in $(\epsilon-\epsilon_0)/\epsilon$ is $(10^4)^{1/50} = 1.202$. On the other hand, when ν and k are kept fixed but the number of oscillators is allowed to change, larger changes in kinetic shift can occur. Thus, when $k = 10^5$, $\nu = 10^{10}$ and $(s-1)$ $= 10$, the relative kinetic shift $= 10^{-\frac{1}{2}} = 0.3162$, but when $(s-1) =$ 50 this is increased to 0.7943. Hence, when the rise of k with ϵ is relatively steep (few effective oscillators), the kinetic shift is smaller.

These criteria provide a rationale for the interlinked factors that influence the kinetic shifts and therefore the appearance potentials of metastable as well as normal fragment ions, and especially account for the empirical observation that relatively strong metastable peaks are usually associated with rearrangement processes involving elimination of stable neutral species, where the frequency factors are low.

Measurement of the appearance potentials of normal fragment ions and metastable ions involving the same reaction mechanism can, in addition, provide useful information on the likely magnitude of the kinetic shift and therefore on the correction that must be made in appearance potential data in order to infer values for activation energy. For example, Hertel and Ottinger [132] find that the appearance potentials of metastable and daughter ions for the reaction involving HCN elimination from the benzonitrile molecular ion differ by as much as 1.3 eV and this shows that the kinetic shift for the metastable is at least as large as this.

The relation between these energies follows from the rate expressions

$$\nu \left(\frac{\epsilon^* - \epsilon_0}{\epsilon^*} \right) = k^*$$

where ϵ^* represents the internal energy of the metastable ion and k^* the rate constant for its fragmentation, and

$$\nu \left(\frac{\epsilon^F - \epsilon_0}{\epsilon^F} \right) = k^F$$

where ϵ^F and k^F refer to the daughter ion. Thus,

$$\left(\frac{\epsilon^* - \epsilon_0}{\epsilon^F - \epsilon_0} \right) = \left(\frac{k^*}{k^F} \right)^{1/(s-1)}$$

or

$$\frac{\epsilon^* - \epsilon_0}{\epsilon^F - \epsilon^*} = \frac{\left(\dfrac{k^*}{k^F}\right)^{1/(s-1)}}{1 - \left(\dfrac{k^*}{k^F}\right)^{1/(s-1)}} = \frac{\text{kinetic shift for metastable ions}}{\text{difference in A.P.}}$$

This may clearly exceed unity for $k^*/k^F \cong 1/10$ and assuming that s may be taken to be about one-fifth of the actual number of oscillators, $i.e.$ about 6 for C_6H_5CN, one deduces that $(\epsilon^* - \epsilon_0)/(\epsilon^F - \epsilon^*)$ = 1.7. Thus the kinetic shift appears as large as 2.2 eV.

SIMPLIFIED APPLICATIONS OF THE QUASI-EQUILIBRIUM THEORY TO COMPLEX ORGANIC COMPOUNDS

Whilst the original theoretical studies based on the quasi-equilibrium theory were necessarily restricted to relatively small molecular species such as propane for which the rate constants were derived on the basis of plausible assumptions concerning frequencies of vibrational and rotational oscillators, it has recently been shown by Williams and coworkers[145,281] that a simplified version of QET may be extended advantageously to a wider range of organic compounds, including systems more directly of interest to the organic chemist. Notwithstanding the rather drastic simplifications, it may reasonably be claimed that all the really essential features are retained and thus the measure of agreement between theory and experiment is valuable evidence in support of the broad assumptions of the theory itself, at least as a didactic framework. This argument cannot be carried too far, however, as there are many precedents for theories which achieve success by all the pragmatic standards and yet may involve certain fortuitous self-cancelling errors. Thus, the demonstration that the theory works as if the electronic ground state alone were implicated in the reaction mechanisms is not conclusive proof that other states are excluded. There is little doubt that the proper tool for specifying the electronic state is not the Dirac density matrix appropriate to a pure quantum state but rather some form of modified canonical density matrix where the electronic energy is not sharply defined. This is still true even though the ion is an isolated system so that Boltzmann factors in the canonical density matrix have not their conventional thermodynamic significance, since the "temperature" of an isolated system lacks definition. But it may be shown [175] that the Dirac density matrix $\rho(r, r_0, \epsilon)$ for a sharply defined electronic energy is merely the Laplace transform of the canonical density matrix $C(r, r_0, \beta)$ for the case of free electron wave functions

$$C(r, r_0, \beta) = \beta \int_0^\infty \exp(-\beta\epsilon)\, \rho(r, r_0, \epsilon)\, d\epsilon$$

There is nothing in the way the QET is formulated to distinguish clearly whether one is dealing with a sharp energy value (such as that for the ground state) or some form of mean energy appropriate to the case of random internal energy transfer by radiationless transitions wherein the Born—Oppenheimer principle may lose validity at every potential surface intersection. Thus the electronic properties could just as well be based on the canonical density matrix as on the sharp energy form. It appears that the necessary *ab initio* calculations to distinguish between the two for the case of ion fragmentation processes have not yet been made.

Yeo and Williams [282] select seven compounds which essentially undergo only two consecutive reactions (exceeding 96% of the total fragmentation) from their molecular ions; earlier applications have been to mono-substituted and di-substituted benzenes where either a single fragmentation process or two competing reactions were considered. A typical example is the consecutive loss of the methoxy radical and CO from methylbenzoate. Rather than attempt to evaluate frequency factors directly, where the lack of precise knowledge would merely be referred to more fundamental parameters, the authors assign values for particular types of fragmentation, for example, 4×10^{13} sec^{-1} for simple bond cleavage, 3×10^{10} sec^{-1} for rearrangement and so on. The rate constant, k, is taken to be given by the simplest QET expression

$$k = \nu \left(\frac{\epsilon - \epsilon_0}{\epsilon} \right)^{(s-1)x}$$

where x measures the fraction of oscillators that are effective, which can be as low as a fifth at the threshold. To make allowance for the greater number of effective oscillators with increasing energy, the explicit assumption is made that

$$x = 0.2 + 0.03\, (\epsilon - \epsilon_0)$$

where the excess energy is measured in volts. (This assumption takes no account of the fact that s varies differently for tight and loose activated complexes.) The linear variation becomes invalid for much greater internal energies where x would exceed unity by an absurd amount. Some assumption needs to be made about the excess energy in the fragment ions. For the first stage, an equipartition assumption implies excess energy $\eta(\epsilon-\epsilon_0')$, where η is the equipartition factor given by the numbers of vibrational degrees of freedom of fragment ion and radical and ϵ_0' is the activation energy required to form the primary product. For the subsequent stage, by formal analogy, it is assumed that the energy in excess of that required to attain the transition state is $\eta(\epsilon-\epsilon_0'')$, although it could be argued that since the intermediate fragment contains less than the original total internal energy, ϵ, this should be reflected in the equipartition formula, viz.

$$\eta[\eta(\epsilon-\epsilon_0')-\epsilon_0'']$$

Finally, a simple modified triangular shape is adopted for the energy distribution curve and the fragmentations occurring within the field-free region are derived by numerical integration for the known times of entering and leaving this region for a particular instrument (the AEI type MS 9). The extent of agreement, which seems very satisfactory, is well represented by the results for the methyl benzoate ion shown in Fig. 99.

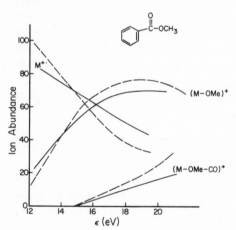

Fig. 99. Application of a simplified form of the quasi-equilibrium theory to the fragmentation of methyl benzoate. The calculated breakdown patterns (broken lines) for the three major ions are compared with the experimental patterns (solid lines).

EQUIPARTITION EFFECTS IN HOMOLOGOUS SERIES

McLafferty and Pike[190] observed that in an homologous series of positive ions fragmenting to a common ion P^+ which itself can undergo further dissociation, $\ln(m*/[P^+])$ is linearly related to the reciprocal of the number of degrees of freedom in the initial molecular ion. Qualitatively, this can be understood as a "degree of freedom effect". For when

$$(RP)^{+\cdot} \rightarrow R^\cdot + P^+$$

the corresponding degrees of freedom are $n \rightarrow (n-s) + s$.

The energy of P^+, assuming equipartition, is $\epsilon s/n$ where ϵ is the excess energy in the original $(RP)^{+\cdot}$ ion. For an homologous series, the excitation energy function is controlled by some most readily ionized group and will not vary to any great extent for different homologs. Thus ϵ is a characteristic excess energy appropriate to all ions and the quantity $\epsilon s/n$ which determines the instability of P^+ varies as the reciprocal of the degrees of freedom n. It is reasonable that $\ln(m*/[P^+])$ provides a measure of this relative instability and the significance of the observation of McLafferty and Pike is then clear.

Lin and Rabinovitch [171] have further substantiated the degrees of freedom effect by a more quantitative treatment of internal energy partitioning. Whilst recognizing that some questionable approximations are necessarily introduced, these authors appear to have been successful in providing a convincing evaluation of the internal energy distribution amongst the fragment species. The commonly adopted approximation of a parabolic internal energy distribution within the molecular ion is introduced. It is further assumed that randomization of the excess energy

$$\epsilon^\ddagger = \epsilon* - \epsilon_0$$

amongst the active degrees of freedom of the activated complex, as in the RRKM formulation, occurs. The reaction scheme for primary alcohols is envisaged as follows.

$$RC_2H_4OH^{+\cdot} \rightarrow C_2H_5O^+ + R^\cdot \xrightarrow{m* = 8.0} H_3O^+ + C_2H_2$$
$$(M^{+\cdot}) \qquad\quad (P^+) \qquad\qquad\qquad\qquad (F^+)$$
$$\xrightarrow{m* = 18.7} CHO^+ + CH_4$$
$$(F'^+)$$

The probability of energy $\Delta\epsilon$ being found in P^+ is considered to be proportional to the product of the degeneracy of internal degrees of freedom of P^+ and the sum of the degeneracies of permitted energy eigenstates of the active degrees of freedom of the rest of the molecular ion at energy $(\epsilon^{\ddagger}-\epsilon)$.

It turns out on the basis of the calculation of the probability that energy $\Delta\epsilon$ should be found in $C_2H_5O^+$, that one might expect the relative abundance of the metastables to vary substantially as the reciprocal of the number of degrees of freedom of the molecular ion, as McLafferty and Pike proposed. The actual theoretical plots on this basis showed some slight curvature, though the essential features of the correlation appear to receive theoretical confirmation.

TUNNEL EFFECTS

If we may suppose that an ion with s effective degrees of freedom and internal energy ϵ may undergo fragmentation when energy (ϵ') less than the barrier height (ϵ_0) is available in the appropriate reaction co-ordinate then, by a generalization of the simple rate expression, we can suppose that

$$k = \nu \left(\frac{\epsilon-\epsilon'}{\epsilon}\right)^{s-1} \exp\left(\frac{-2\pi(\epsilon_0-\epsilon')}{h\nu_t}\right)$$

The tunneling factor is given by Bell[24] for a parabolic barrier and $\nu_t = \epsilon_0^{\frac{1}{2}} / \{\pi a(2m)^{\frac{1}{2}}\}$ where m is the effective mass and $2a$ the barrier width at half height.

There are two difficulties in such a generalization. First, the frequency factor, ν, to be used when barrier penetration occurs, may not be the same when the dissociation occurs in the classical manner. Secondly, by the QET hypothesis of parallel reactions all possible barrier penetration energies ϵ' should be considered and it may well be that the classical process with $\epsilon' \equiv \epsilon_0$ is a faster one than any of these. For example, it may be that k for the normal reaction is 10^8 sec^{-1} when ϵ takes an appropriate value, whereas k for some $\epsilon' < \epsilon_0$ is 10^6 sec^{-1}. Whilst it is true that the latter process leads to a metastable ion, the faster process will have reduced the ion concentration in 10^{-6} sec by a factor of $e^{-kt} = e^{-100}$! The required condition that tunneling shall be a preferred process is therefore that k shall have a sharp maximum at some $\epsilon'/\epsilon_0 < 1$ for ions of a certain internal energy ϵ.

From

$$\ln k = \ln \nu + (s-1)\ln \left(\frac{\epsilon-\epsilon'}{\epsilon}\right) - \frac{2\pi(\epsilon_0-\epsilon')}{h\nu_t'}$$

$$\left[\frac{\partial(\ln k)}{\partial\epsilon'}\right]_{\epsilon,\,\epsilon_0,\,\nu,\,s} = -\frac{(s-1)}{(\epsilon-\epsilon')} + \frac{2\pi}{h\nu_t'}$$

whence

$$\epsilon-\epsilon' = \frac{(s-1)h\nu_t'}{2\pi}$$

Comparing the barrier penetration height for two similar processes

$$\epsilon' = \epsilon - \frac{(s-1)h\nu_t'}{2\pi}$$

$$\epsilon'' = \epsilon - \frac{(s-1)h\nu_t''}{2\pi}$$

or $$\epsilon'' - \epsilon' = \frac{(s-1)}{2\pi}h(\nu_t'-\nu_t'')$$

It may be assumed that the energy in the reaction co-ordinate itself at the instant of dissociation determines the kinetic energy liberated. This is, however, not always the case as discussed in Chapter 4. On this assumption, however, it follows that the process where the effective mass in the dissociation co-ordinate is greater will liberate the greater amount of kinetic energy. Thus, a comparison of exothermicities for the fragment ions $CH_2=\overset{+}{O}H$ or $CH_2=\overset{+}{O}D$ to form HCO^+ indicated a greater release of energy from the deuteriated ion (by approximately 0.22 eV). This was confirmed by similar comparisons of $CH_2=\overset{+}{O}D$ (ref. 39). It seems probable that the negative curvature at the saddle point is responsible for the isotopic inversion effect, for the normally expected result of substituting a higher mass isotope would be to reduce the zero point energy and the accompanying energy release when the bond dissociates.

The significance of the frequency factor may perhaps be clarified if the simplest case of a plane wave e^{ikx} impinging on a simple rectangular energy barrier of finite width a and height ϵ_b is considered[210]. The flux of particles, j, when the barrier is absent is given by

$$j = \frac{\hbar}{2mi} (\overline{u} \text{ grad } u - u \text{ grad } \overline{u})$$

where $u = e^{ikx}$ and thus, $j = \hbar k/m$.

If T_c is the transmission coefficient for barrier penetration when the kinetic energy on the approach side of the barrier is less than ϵ_b then

$$j = \frac{\hbar k T_c}{m}$$

Identifying $\hbar k/ma$ formally with the frequency factor, we may write $j = \nu a T_c$ and it may be assumed that for other barrier shapes the classical frequency factor may be retained in the rate expression provided the correctly evaluated form for the transmission coefficient is included as a factor. This is the justification for our use of the simple modification of the QET rate expression.

Current investigations are bringing to light much new and interesting information on isotope effects in various types of metastable ions. It seems likely that in the near future it will be possible to say how the isotope effect upon the kinetic energy release for such ions depends on specific mechanistic and structural features[31, 54, 69, 170].

It seems generally true that, in the case of one-step hydrogen loss mechanisms of a simple cleavage type, T_H/T_D is less than unity but not much less[31] except perhaps for special cases such as that first encountered in methanol[39]. This reaction involves further dissociation of a daughter ion

$$CH_2O^{+\cdot} \rightarrow CHO^+ + H^{\cdot}$$

and T_H/T_D is much lower (0.48) than typical values for molecular ion dissociation processes. No convincing explanation of this abnormally low value, without invoking some rather special feature such as a tunneling mechanism, seems to have so far been propounded. But it is not suggested that recourse to this hypothesis is always necessary.

Certain qualitative suggestions can, however, be made. Both the reverse activation energy ($\epsilon_0^{\ r}$) and the internal energy of the activated complex ($\epsilon^{\ddagger}$) can contribute to the kinetic energy release as is illustrated in Fig. 100. Where $\epsilon_0^{\ r}$ is considerable, it dominates the situation because only a small fraction of $\epsilon^{\ddagger}$ is able to enter the

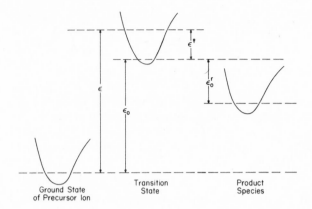

Ground State
of Precursor Ion   Transition
                          State   Product
                                         Species

Fig. 100. Excess energy terms which may contribute to the kinetic energy of fragment separation.

dissociation co-ordinate in any case. Moreover, $\epsilon^{\ddagger}$ for metastable ions is substantially the energy quantity which we have previously referred to as the kinetic shift, because it is the energy excess $(\epsilon - \epsilon_0)$ over and above the minimum potential energy required to negotiate the energy barrier. As may be appreciated from consideration of the simple rate expression and as we have emphasized earlier, for polyatomic ions with many internal degrees of freedom the $\epsilon^{\ddagger}$ contribution becomes important. For a particular case with $(s-1) = 10$, we calculated a value of 0.3162 whereas for an ion about five times as large, $(s-1) = 50$, in the same units the energy excess had risen to 0.7943. But on a basis of equipartition, the fraction of this energy which can enter the dissociation co-ordinate and contribute to kinetic energy varies substantially as $1/s$, which factor more than compensates for the higher total internal energy in the larger ion.

Again, if ν_1, ν_2, ν_3 are frequencies of the stretching mode that becomes the dissociation degree of freedom and two accompanying bending mode frequencies for the R—H bond, the frequency factor for the hydrogen elimination can be written approximately in the form

$$\nu_1 \nu_2 \nu_3 / \left(\nu_2^{\ddagger} \nu_3^{\ddagger} \right)$$

where $\nu_2^{\ddagger}$ and $\nu_3^{\ddagger}$ refer to activated state frequencies. We assume for ease of exposition that other frequencies in the initial and activated states are self-canceling.

In those cases where $\nu_2 \cong \nu_2^{\ddagger}$ and $\nu_3 \cong \nu_3^{\ddagger}$, we are left simply with a frequency factor for the reaction which is the same as the frequency of the stretching mode of vibration in the stable form of the ion. If hydrogen is replaced by deuterium, the frequency is reduced by a factor very close to $(0.5)^{1/2}$. Again, on consideration of the rate expression, ϵ must be increased to maintain the same rate as for hydrogen elimination. Because not more than about ϵ/s can contribute to kinetic energy, we naturally expect T_H/T_D to be less than unity but not very much less. This seems to be typically the result established for "normal" metastable dissociations involving molecular ions. In cases where the kinetic energy release is large, $\epsilon_0{}^r$ must be large and in this case $T_H \cong T_D$ if the dissociation proceeds classically. However, if tunneling occurs, T_H/T_D values below unity may result even in the case where $\epsilon^{\ddagger} \ll \epsilon_0{}^r$.

Other interesting situations can arise according as the bending mode frequencies enter into the argument. They will clearly not cancel exactly in the frequency factor expression. It is not possible to enter into all the detailed effects that can be expected according to the interlinked ways in which zero point energy components can change the situation and the correlated changes in ϵ and $\epsilon_0{}^r$. But at least the broad features of the effects observed are gradually receiving interpretation.

From the experimental standpoint, it is often easier to interpret results where two functionalities, R—H and R—D, are present in the same species, for in that case at least we know that the total internal energy is the same and this reduces the number of variables to be considered in discussing the competitive processes. Where different species are considered, they will receive different degrees of excitation to attain the required order of reaction rate ($10^5-10^6\,\text{sec}^{-1}$). For the latter it is often not possible to say more than $T_H/T_D \cong 1$.

RATES OF CROSSING OF ENERGY SURFACES AND SIMILARITIES TO TUNNELING ASPECTS

Closely related to the tunneling mechanism is that of crossing from one potential energy surface to another. If the surfaces derive from states of the same symmetry type, perturbation at the point of crossing is strong and it is improbable that a particle moving over one energy surface will ever have sufficient energy to surmount this energy barrier and to cross over on to the other. Such perturbations may, however, be restricted to a fairly narrow region around the cross-over point and then an effect essentially of the tunneling type can arise, namely that the kinetic energy of the nuclei may be insufficient to cross over but sufficient to tunnel through the barrier. It will be quite difficult to determine whether tunneling is of the simple type involving a single potential energy surface with an activated region or involves two surfaces in close proximity. Again, if the surfaces correspond to different symmetry species, perturbation energies will be of small amplitude and breadth. If one surface permits continuum wave functions, then this presents what is, in effect, an energy barrier that is readily tunneled through by a dissociating ion because the intersection is equivalent to a cusp which has the ideal sharp curvature to facilitate tunneling. In fact, it may well be that the sharp peaks in energy barriers most likely to lead to tunneling phenomena can be most readily accounted for if they originate from potential surface intersections.

Coulson and Zalewski[92] have given a fairly exhaustive treatment of such phenomena and evaluated transition moments for various combinations of an initially bound or continuum state proceeding to a final state of either type. Whilst differing in detail, these authors' expression for the probability of a transition is still, for most purposes, substantially similar to the original expression of Zener[284] and Landau[167], viz.

$$p = \exp \frac{-4\pi^2 \, \epsilon_{12}{}^2}{h\upsilon \, |s_1 - s_2|}$$

where ϵ_{12} is the perturbation energy, υ the velocity at the point of crossing and $|s_1 - s_2|$ is the difference of the slopes of the intersecting potential surfaces. This latter measures, in some way, the degree of sharpness of the cusp given by the combined surface. If it is very sharp, the probability of crossing approaches unity for sufficiently large velocities.

The tunneling factor is given by[24]

$$\exp \frac{-2\pi(\epsilon_0 - \epsilon')}{h\upsilon_t{}'}$$

Again $\upsilon_t{}' = \epsilon_0{}^{1/2}/\{\pi a(2m)^{1/2}\}$ measures the sharpness of the top of the barrier, becoming large when a is small, and fulfilling the same function as the cusp sharpness $|s_1 - s_2|$ in the Zener—Landau expression.

Further work will be necessary to determine whether certain curious effects of metastable ion dissociation may be best explained by simple tunneling, cross-over of energy surfaces, or in other ways.

POTENTIAL ENERGY SURFACES AND CORRELATION RULES

Whereas it is usually difficult to determine the detailed behavior of potential energy surfaces, for small molecular ions of high symmetry, correlation effects between initial states and those of dissociation products enable useful and informative conclusions to be drawn. This has become a fashionable area for investigation in recent years, largely associated with the important advances in understanding of thermal and photochemical reactions provided by the Woodward—Hoffman rules[279]. Earlier, the pioneering work of Walsh[266, 267] laid the foundation for theoretical understanding of molecular geometry and symmetry. McDowell[182] has also extensively used group theoretical approaches to dissociation processes in mass spectrometry and the insight provided by the Wigner—Witmer correlation rules was demonstrated, in particular by Laidler[166].

· From the investigation of metastable ions originating from fragment ions such as CH_2OH^+ in methanol, it has been shown that some features of QET and specific quantum chemical effects such as tunneling and potential surface crossing may be combined. Fiquet-Fayard and Guyon[110] have, however, recently argued that QET may perhaps have no application to mechanisms including surface crossing and predissociation, for in these cases a sharply defined energy is necessary. Unlike the QET case where a distribution of energies of arbitrary value exceeding the appearance potential of a particular fragment is involved, too much energy will impede, if not preclude, pre-dissociation because motion of the effective mass continues beyond the point of crossing undisturbed by the small perturbation energy at this point. Clearly such mechanisms are more sensitive to isotopic effects than are those of QET which involves no specific energy properties.

Fiquet-Fayard discussed the potential surfaces $^2B_1$, $^2A_1$, $^2B_2$ of the $H_2O^{+\cdot}$ ion as already discussed by Al-Joboury and Turner[2]. Using the correlation rules, it may be shown that the ground state of OH^+ ($^3\Sigma^-$) does not correlate with these initial states. $^2B_2$ would give $^1\Delta$ (OH^+) by correlation rules. Production of $^3\Sigma^-$ fragment ions is therefore evidence for predissociation.

The general principle of correlation is illustrated by the hypothetical formation of $H_2O^{+\cdot}$ from H_2 in the ground state ($^1\Sigma_g^+$) and the oxygen cation in the ^2D_u state as shown in Fig. 101. It is supposed that the reaction proceeds by approach of $O^{+\cdot}$ along a symmetry axis in which C_{2v} symmetry is preserved throughout. The character table for the C_{2v} group is as follows.

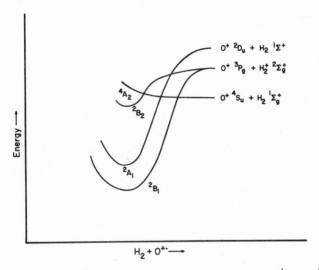

Fig. 101. Correlation diagram for the reaction $H_2O^{+\cdot} \rightarrow O^{+\cdot} + H_2$.

		E	C_2	σ_v	$\sigma_{v'}$
A_1	z^2	1	1	1	1
	x^2-y^2				
A_2	xy	1	1	-1	-1
B_1	xz	1	-1	-1	1
B_2	yz	1	-1	1	-1

Clearly the $O^{+\cdot\,2}D_u$ state contains irreducible representations transforming as z^2, xy, xz, zy, x^2-y^2 under C_{2v} whilst $^1\Sigma_g^+$ transforms as A_1.

Thus $^2D_u + {}^1\Sigma_g^+ \rightarrow {}^2A_2 + {}^2B_2 + {}^2B_1 + {}^2A_1$

Consider the correlation with the C_s group in splitting off a hydrogen atom,

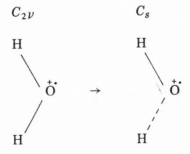

and the character table for C_s, namely

	E	$\bar{\sigma}_h$
A'	1	1
A''	$\cdot 1$	-1

In view of the $^3\Sigma^-$ state transforming like the A'' representation, it follows that only an A'' representation can correlate directly with the low-lying states corresponding to removal of the three least bonding electrons in H_2O. This provides evidence for predissociation involving $^4A_2$, $^4A''$. Because of the change of spin, the perturbation between the two states is very slight and provides a very sharp curvature in the cross-over region. Isotope effects such as tunneling will be very pronounced because of this. The transition integral involves the spin-orbital part of the Hamiltonian,

$$A(S \cdot L) = A(S_x L_x + S_y L_y + S_z L_z)$$

where

$$A = \frac{1}{2m^2c^2} \sum_k \frac{1}{r} \frac{\partial U(r)}{\partial r}$$

and $U(r)$ is the potential field of any nucleus.

Whilst, in principle, the motion of one electron relative to another implies an additional magnetic field, the Coulombic repulsion of like charges prevents the electrons from approaching close enough for the coupling coefficient to become appreciable. Thus, it appears that electron—nuclear interactions alone are primarily responsible for changes in spin multiplicity and predissociation.

The rules for correlation of initial and product states may be generalized. The general procedure involves resolving the direct product of the initial states into symmetry species appropriate to the geometry of the reaction complex. The only theoretically possible reaction paths are those where the same symmetry type is determined by the direct product of the product states. The spin multiplicity of the reaction complex is determined by the vector addition rule for angular momenta.

In the case of predissociation a single surface resulting from spin—orbital interaction terminates in different spin multiplicities. Such a reaction would not strictly be describable as an "adiabatic" process as defined by the Wigner—Witmer rules, but this term is sometimes used to describe any reaction which takes place over a single, continuous potential energy surface.

MATHEMATICAL SECTION

The RRKM evaluation of internal energy distributions

In this book we use extensively rate expressions for the dissociation of positive ions derived from statistical considerations. It does not matter so much which type of mathematical approach to the statistical properties we prefer to adopt since all lead to essentially similar conclusions. The most important approaches have been those of Wahrhaftig, Rosenstock and coworkers (the so-called QET method)[231], the earlier type of formulation due to Rice, Rampsperger, Kassel and Marcus (referred to commonly as the RRKM theory)[158, 176, 228] and the rather different theory due to Slater[246]. Here we summarize the main results of the RRKM theory, as far as our own applications are concerned. A summary of the equivalent QET equations is appended so that it should be possible to see at a glance the relationship between the two main approaches.

Consider the case of n quanta to be assigned to s oscillators. The number of assignments is the number of ways of arranging one group of $(s-1)$ identical objects (essentially partitions) with n others (vibrational energy quanta). According to the elementary algebra of partitions, the required result is

$$\frac{(n + s - 1)!}{n! \ (s - 1)!} = \frac{(n + s - 1) \ (n + s - 2) \ldots \ldots (n + 1)}{(s - 1) \ (s - 2) \ldots \ldots 1}$$

If $n \gg s$ this reduces, to a very good approximation, to

$$\frac{n^{s - 1}}{(s - 1)!}$$

Suppose, now, that the activated state requires a minimum of r quanta localized in the reaction co-ordinate before it is able to dissociate. That is, $\epsilon_0 = rh\bar{\nu}$ where ϵ_0 is the energy of activation. Any ion with $r, r + 1, r + 2, \ldots$ quanta will be able to dissociate, though the number of ions with more than just sufficient quanta will fall away rapidly and contribute relatively little to the quota of dissociating ions.

The number of assignments of $(n-r-u)$ quanta amongst the $(s-1)$

stable co-ordinates is clearly $(n-r-u)^{s-2}/(s-2)!$ and the total number is therefore given by

$$\sum_{u=1}^{n-r} \frac{(n-r-u)^{s-2}}{(s-2)!}$$

Now

$$\sum_{u=1}^{n} u^p = \frac{n^{p+1}}{p+1} + \frac{n^p}{2} + \frac{B_1}{2!}\,pu^{p-1} + \ldots\ldots$$

where B_1 is the first Bernoulli number [103]. For large n, all terms after the first one are relatively small and may be neglected. The required summation is thus seen to have the value

$$\frac{(n-r-1)^{s-1}}{(s-2)!\ (s-1)} = \frac{(n-r-1)^{s-1}}{(s-1)!}$$

The fraction of the total number of ions in the activated state

$$= \frac{(n-r)^{s-1}/(s-1)!}{n^{s-1}/(s-1)!}$$

$$= [(n-r)/n]^{s-1}$$

$$= \left(\frac{\epsilon-\epsilon_0}{\epsilon}\right)^{s-1}$$

This provides a statistical evaluation of the fraction of ions at any one time having at least sufficient energy to decompose. Suppose that it requires an average time τ for an activated ion to dissociate. Then the number of ions dissociating per unit time is

$$\tau^{-1}\left(\frac{\epsilon-\epsilon_0}{\epsilon}\right)^{s-1}$$

and τ may be equated to a mean frequency ν of crossing the energy barrier. The proper value to be assigned to $\bar{\nu}$ emerges from detailed consideration of motion of an equivalent mass along the reaction co-ordinate. The appropriate way of defining the mean frequency is best given by the methods of the QET approach which follows.

Some formal properties of the quasi-equilibrium theory

With the background of qualitative aspects, it is now possible to formalize some of these concepts and to arrive at the simplest forms in terms of a particular oscillator model which is widely employed in practice[2][3][1].

In the energy range ϵ to $(\epsilon + d\epsilon)$, the ion undergoing fragmentation may be supposed to have $\rho(\epsilon)d\epsilon$ accessible states where $\rho(\epsilon)$ is a density function. Similarly, the activated complex with potential energy ϵ_0 and kinetic energy between ϵ_t and $\epsilon_t + \delta\epsilon_t$ localized in the reaction co-ordinate has $\rho^{\ddagger}[(\epsilon,\epsilon_0,\epsilon_t)\rho_t(\epsilon_t)]\delta\epsilon\delta\epsilon_t$ accessible states. Let $r(\epsilon_t)$ be the reaction rate for the case where kinetic energy ϵ_t is present in the reaction co-ordinate. By assumption of equal *a priori* probabilities for all accessible states, the weighting factor for the rate $r(\epsilon_t)$ is

$$\frac{\rho^{\ddagger}(\epsilon,\epsilon_0,\epsilon_t)\rho_t(\epsilon_t)}{\rho(\epsilon)}$$

and the overall rate of reaction is given by

$$k = \int_0^{\epsilon-\epsilon_0} \frac{\rho^{\ddagger}(\epsilon,\epsilon_0,\epsilon_t)\,\rho\,(\epsilon_t)\,r\,(\epsilon_t)\,\delta\epsilon_t}{\rho(\epsilon)}$$

The reaction co-ordinate may be discussed in terms of a one-dimensional box of length l situated at the top of the energy barrier where the barrier is sensibily flat if we exclude very sharply curved barriers readily penetrated by quantum mechanical tunneling. This refinement is discussed separately in connection with some isotopic inversion phenomena.

Using the familiar result for energy levels ϵ_t in a one-dimensional box

$$\epsilon_t = \frac{n_t^2 h^2}{8\mu l^2}$$

where n_t is the translational quantum number, l a hypothetical distance along the reaction co-ordinate and μ the "reduced mass" for motion in the reaction co-ordinate. Then

$$\rho_t(\epsilon_t) = \frac{dn_t}{d\epsilon_t} = \frac{1}{h(2\mu/\epsilon_t)^{1/2}}$$

and the rate of traversal through the saddle point in the forward direction is

$$\frac{1}{2l}\left(\frac{2\epsilon_t}{\mu}\right)^{1/2}$$

since the velocity, v, is given by $\frac{1}{2}\mu v^2 = \epsilon_t$.

The rate of dissociation is then given by

$$k = \int_0^{\epsilon-\epsilon_t=0} \frac{\rho^{\ddagger}(\epsilon,\epsilon_0,\epsilon_t)}{\rho(\epsilon)} \frac{1}{h} \delta\epsilon_t$$

If it is now supposed that the internal energy is distributed amongst s oscillators of equal classical frequency, v, then each oscillator contributes energy $(n_i + \frac{1}{2})hv$ so that

$$\sum_{i=i}^{s} (n_i + \frac{1}{2})hv = \epsilon$$

or

$$\sum_{i=i}^{s} n_i = \left[\frac{\epsilon}{hv} - \frac{s(s+1)}{4}\right] = N$$

The number of ways of partitioning the energy into s packets of energy is the same as the number of ways of partitioning N objects and is therefore given by elementary algebra as

$$\frac{(N+s-1)!}{N! \ (s-1)!}$$

Similarly, for the activated complex the number of partitions is

$$\frac{(N^\ddagger + s-2)!}{N^\ddagger!(s-2)!}$$

where

$$\left[\frac{\epsilon-\epsilon_0-\epsilon_t}{h\nu} - \frac{s(s-1)}{4}\right] = N^\ddagger$$

Now

$$\frac{(N+s-1)!}{N!\,(s-1)!} \cong \frac{N^{s-1}}{(s-1)!} \quad \text{if } N \gg s$$

It will be clear that this approximation is not closely obeyed when N is not large and for threshold energies the retention of the same formal expressions has the result that s must be reduced by a factor of about 3 or 5 to achieve a more realistic energy dependence.

Thus

$$\frac{\rho^\ddagger(\epsilon,\epsilon_0,\epsilon_t)}{\rho(\epsilon)} = \frac{N^{\ddagger\,s-2}/(s-2)!}{N^{s-1}/(s-1)!}$$

$$= (s-1)\frac{N^{\ddagger\,s-2}}{N^{s-1}} \cong \frac{(s-1)h\nu(\epsilon-\epsilon_0-\epsilon_t)^{s-2}}{\epsilon^{s-1}}$$

whence

$$k = \frac{(s-1)\nu}{\epsilon^{s-1}} \int_0^{\epsilon-\epsilon_0} (\epsilon-\epsilon_0-\epsilon_t)^{s-2}\,\delta\epsilon_t$$

$$= \nu\left(\frac{\epsilon-\epsilon_0}{\epsilon}\right)^{s-1}$$

Equivalent, but more complicated, derivations are given in the literature, taking into account differences in the frequencies for the harmonic oscillations in the molecular ion and the activated state. Instead of the simple frequency factor, ν, one then obtains a certain average frequency given by

$$\frac{\prod\limits_{i=1}^{s} \nu_i}{\prod\limits_{i=1}^{s-1} \nu_i}$$

A similar analysis including free rotors of energy ϵ_j given by

$$\epsilon_j = \frac{n_j^2 h^2}{8\pi^2 I_j}$$

(where n_j is the rotational quantum number and I_j the reduced moment of inertia) leads to a final rate expression where half integer powers of L, the number of free internal rotors, appear on account of $\epsilon_j^{1/2}$ appearing in

$$\frac{d\epsilon_j}{dn_j} = \frac{h^2 \epsilon_j}{2\pi^2 I_j}$$

Kropf *et al.* [164] give a corrected version of the final rate expression, which has the form

$$k = \nu \left(\frac{\epsilon - \epsilon_0}{\epsilon}\right)^{N-(L/2)-1} (\epsilon - \epsilon_0)^{(L-L^{\ddagger})/2}$$

and the frequency factor, ν, now includes numbers and moments of inertia of internal rotors, *viz*:

$$\nu = \sigma \, (2\pi)^{3/2(L-L^{\ddagger})} \; \frac{\Gamma(s - L/2) \prod\limits_{f=1}^{L^{\ddagger}} \left(\frac{I_f^{\ddagger}}{n_i^{\ddagger}}\right)^{1/2} \prod\limits_{j=(L+1)}^{s} \nu_j}{\Gamma(s - L^{\ddagger}/2) \prod\limits_{i=1}^{L} I_i^{1/2} \prod\limits_{g=(L^{\ddagger}+1)}^{(s-1)} \nu_g^{\ddagger}}$$

For semiquantitative purposes, which generally suffice to clarify the salient features of energy dependence of fragmentation mechanisms, the simple harmonic oscillator form derived above is still useful, since the incidence of rotors merely introduces a further refinement into the definition of the effective number of oscillators. Whether these occur or not, for threshold energies as in the case of metastable ions, there are other approximations in the derivation which are apparent in the simplified version given here and the number of effective oscillators has always, at best, the nature of an empirical parameter.

The density of states formulae are approximate since the number of ways of partitioning the energy is

$$\frac{(n+s-1)!}{n! \, (s-1)!} = \frac{1}{(s-1)!} \prod_{i=1}^{s-1} (n+i)$$

and the classical result corresponds to the replacement of each term in the product of n. Other approximations have been suggested[2][3][5]. One of these, the so-called semi-classical approximation, uses the average term

$$n + \frac{\left(\sum\limits_{i=1}^{s-1} i\right)}{(s-1)} = n + \tfrac{1}{2}s$$

when the classical rate formula $n^{s-1}/(s-1)!$ is replaced by $(n+\tfrac{1}{2}s)^{s-1}/(s-1)!$. Since $\epsilon_j = n_j h\nu$ (classical), and $\epsilon_j = (n_j + \tfrac{1}{2})h\nu$ (quantum mechanical), the modified result corresponds to adding on the zero-point energy to the non-fixed energy. Thus, the rather arbitrary procedure of replacing

$$\frac{\epsilon^{s-1}}{(s-1)!} \quad \text{by} \quad \frac{(\epsilon + \tfrac{1}{2}sh\nu)^{s-1}}{(s-1)!}$$

receives justification for threshold energies where n is not very much greater than the number of oscillators. Moreover, there will be a limiting finite rate rather than a vanishing one as the non-fixed internal energy approaches zero.

Generalized wave functions and effects of energy barriers

In general, it may be concluded that the manifold pathways to fragmentation depend on the initial preparation by electron impact of the positive ion in a mixture of eigenstates. From then on, when the ion drifts away from the electron beam, it constitutes an isolated system with an invariant Hamiltonian operator provided we include the nuclear co-ordinates as independent variables as well as those of the electrons. If the Born—Oppenheimer approximation is used, the electronic Hamiltonian is, of course, time-dependent because the potential of the nuclear framework is varying with time as the atoms adopt new configurations as the reaction(s) proceeds. We shall therefore adopt the generalized eigenfunction description $\psi_s(r_1, r_2, \ldots r_m, R_1, R_2, \ldots R_n)$ for a system containing m electrons and n nuclei. If the system were initially prepared in such a state, it would remain in it permanently because the total Hamiltonian is invariant. Thus it may exist as a system which is not well-defined energetically and will have the general form

$$a_s \psi_s(r,R)\, e^{-it\epsilon_s/\hbar}$$
$$(s)$$

where a_s are complex amplitude factors. This time-dependent wave function can represent various configurations of the nuclei coordinates depending on the phase relationships. The initial phase relationships define a highly compressed wave-packet. Subsequently, other much more probable configurations can occur and these correspond to the various fragmentations. Other things being equal, when the time available for reaction permits configurations corresponding to various admixtures of vibrational and kinetic energy, there will be a tendency for energy to be preferentially released in the latter form. The reason for this is that the energy quanta for kinetic energy are much smaller than those for vibrational energy and so a given spread of energy in the kinetic form can permit a greater number of configurations. But this tendency is subject to the limitation that if all the energy were released as kinetic energy with a specific sharp value, the number of configurations would be limited to no more than an arbitrariness in direction (and in separation of the fragments in a classical configuration). If some, but not all, of the energy is vibrational, then not only the direction of motion but the scalar velocity is assignable in many different ways. Thus the number of configurations can be optimized by an energy distribution which is neither wholly vibrational nor wholly kinetic. In the wave mechanical description, the phase relationships will correspond to a release of energy largely, but not entirely, kinetic.

Consider a simple model of an electron moving in a potential field which may be taken to be that of the hydrogen molecule ion. This somewhat simplified model well illustrates the meaning of phase relationships for dynamical processes.

The $1s$ orbitals around each nucleus ψ_A and ψ_B define two approximations to eigenstates.

$$\frac{1}{2^{\frac{1}{2}}}(\psi_A + \psi_B) \qquad \text{(energy } E_1 \text{)}$$

$$\frac{1}{2^{\frac{1}{2}}}(\psi_A + \psi_B) \qquad \text{(energy } E_2 \text{)}$$

If the electron at the time of ionization of H_2 to produce $H_2^{+\cdot}$ is left in one of these eigenstates, it will remain permanently in this state. If, however, it has been localized as a wave-packet on nucleus A, the time-dependent wave function is

$$\overline{\psi}(t) = \frac{1}{2^{\frac{1}{2}}}(\psi_A + \psi_B)e^{-itE_1/\hbar} + \frac{1}{2^{\frac{1}{2}}}(\psi_A - \psi_B)e^{-itE_2/\hbar}$$

which at $t = 0$ reduces to ψ_A.

At some future time, the phases for each eigenfunction change in such a way the $\overline{\psi}(t)$ has the configuration ψ_B. This occurs when

$$\frac{e^{-itE_2/\hbar}}{e^{-itE_1/\hbar}} = -1 = e^{i\pi}$$

which implies that $(E_2 - E_1)t/\hbar = \pi$. The time of "reaction" is $h/2(E_2 - E_1)$.

Now imagine that an energy barrier $V(x)$ is interposed between the nuclei A and B. The energy E_1 is increased by $\int \psi_A V \psi_B dV$ and E_2 is reduced by the same amount. Thus the time required to achieve the correct phase relationships corresponding to "reaction" is increased to

$$\frac{h}{2(E_2 - E_1 - 2)\int \psi_A V \psi_B dV}$$

The nature of the dependence of fragmentation rate processes on the shape and amplitude of the energy barriers is therefore clear. Since the quantity $\int \psi_A V \psi_B dV$ depends on the width as well as the amplitude of the energy barrier, so does the time required to penetrate it. Moreover, according to this model, neither the initial state of the ion immediately after electron impact nor any of its subsequent forms represents an eigenstate of the whole system but rather a particular time-dependent configuration of $\underline{\ddot{\psi}}(t)$.

The energy levels of large polyatomic ions

Consider a positive ion containing an even number of atoms and an odd number of electrons (one less than the number of electrons in the molecule). Let these electrons, n in number, be allocated to a set of distinct atomic orbitals a_1, a_2, a_3, ... a_n. Consideration of determinantal wave functions ϕ:

$$\frac{1}{n^{1/2}} |(a_1 \alpha)_1 \ (a_2 \alpha)_2 \ (a_3 \alpha)_3 \ \ldots \ (a_n \alpha)_n |$$

shows that 2^n such wave functions may be constructed since either of the two spin functions α and β can be assigned to each atomic orbital. Many of these will have the same energy because they belong to the same multiplet. To arrive at the number of distinct energy levels, only one component of each multiplet must be included.

Each multiplet contains one component having one half unit of resultant total spin in the z direction (*i.e.* $S_z \phi = \frac{1}{2} \hbar \phi$).

The number of such determinantal wave functions is therefore given by

$$^n C_{(n-1)/2} = \frac{n!}{\dfrac{n+1}{2}! \ \dfrac{n-1}{2}!}$$

since $\frac{1}{2}(n+1)$ spins α and $\frac{1}{2}(n-1)$ spins β must be assigned to the atomic orbitals to produce the equivalent of a single α spin. For large n values, the number approaches asymptotically $2^{n+1}/n^{1/2}$ which, though small compared with the number of possible wave functions 2^n, is still a very large number. This demonstrates that the potential energy surfaces for large polyatomic ions must be very closely spaced. It should be pointed out, however, that the same atomic orbital may be used twice over provided different spins are assigned. This would reduce further the number of different energy levels so the number evaluated has to be considered an upper limit. By this argument we are also only able to demonstrate the smallness of the average spacing between the potential energy surfaces. The first few energy levels will show much wider spacing and close to the appearance potential of the ion, only these will be involved.

Randomization of isotopic labels

The elimination of a particular fragment when an ion decomposes may be accompanied by exchange of atoms between certain groups. This commonly occurs when groups of atoms become associated in the transition state or when, perhaps due to rotation of side-chains in the ion, atoms can be transferred between remote locations. An example is the fragmentation of benzoic acid where the *ortho* hydrogens can each be exchanged with the hydrogen of the carboxylic acid group, and indirectly with each other, by processes such as[3][2]

Using isotopically labeled compounds, it is possible to study the above exchange and to determine how many such steps precede fragmentation by loss of a neutral hydroxyl radical. As the side chain rotates it is possible to form a series of compounds such as

The degree of labeling of the hydrogen in the hydroxyl group eliminated can be used to estimate the number of exchanges that precede fragmentation.

Isotopic labeling can also be used to study other similar processes such as the exchange reactions in the fragment ion in ketones formed from the molecular ion by a McLafferty rearrangement. This fragment ion can itself lose the elements of a neutral olefin. The use of appropriately labeled compounds can enable deductions to be made concerning which hydrogen atoms are involved in exchange reactions such as

This appendix contains the results of calculations concerned with fragmentations taking place following a very large number of exchanges of the types shown above. The results are given for cases in which x atoms of an isotope A and y atoms of an isotope B are involved in the exchange reactions and become completely scrambled before a fragmentation process occurs in which a total of z atoms of A and B are lost.

In the case of α-d_1-benzoic acid, illustrated above, in which the exchange reactions involve the two *ortho* hydrogens as well as the deuterium atom, we are dealing with two atoms of A and one of B (3 atoms scrambling, formula A_2B) and the table shows that when one atom of either A or B is lost (in the hydroxyl group) the isotopic composition of the atom lost is 0.6667A to 0.3333B. Similarly, in the ketone case, a total of 6 atoms takes part in the scrambling and if the terminal methyl group were labeled with 3 deuterium atoms, we should be dealing with the case of 6 atoms scrambling, a formula A_3B_3 with $z = 2$ when two of these atoms are lost in the neutral olefin.

If the percentage labeling measured in an actual experiment differs from that shown in the appendix, this will be an indication of the operation of a significant primary or secondary isotope effect or that only a limited number of exchanges have taken place before fragmentation, perhaps due to an unfavorable rate constant in the exchange process; the figures in the table will only agree with experimental results when scrambling is complete and no significant isotope factor operates.

The number of atoms scrambling, $(x + y)$, and the separate values of x and y are shown below. This information is listed in the subheadings. The number of atoms, z, lost in the fragmentation is listed in the first column and the corresponding probabilities of all the species $A_nB_{(z-n)}$ in the other columns. Values of $(x + y)$ from 2 to 10 are included; if a combination A_xB_y is listed, the complementary combination A_yB_x is not considered.

Number of atoms scrambling : 2
Formula: AB
1   0.5000 A   0.5000 B

Number of atoms scrambling : 3
Formula: A_2B
1   0.6667 A   0.3333 B
2   0.3333 A_2 0.6667 AB

Number of atoms scrambling : 4
Formula A_3B
1   0.7500 A   0.2500 B
2   0.5000 A_2 0.5000 AB
3   0.2500 A_3 0.7500 A_2B

Formula : A_2B_2
1   0.5000 A   0.5000 B
2   0.1667 A_2 0.6667 AB   0.1667 B_2
3   0.5000 A_2B 0.5000 AB_2

Number of atoms scrambling : 5

Formula: A_4B

1	0.8000 A	0.2000 B
2	0.6000 A_2	0.4000 AB
3	0.4000 A_3	0.6000 A_2B
4	0.2000 A_4	0.8000 A_3B

Formula: A_3B_2

1	0.6000 A	0.4000 B	0.1000 B_2
2	0.3000 A_2	0.6000 AB	0.3000 AB_2
3	0.1000 A_3	0.6000 A_2B	0.6000 A_2B_2.
4		0.4000 A_3B	

Number of atoms scrambling : 6

Formula: A_5B

1	0.8333 A	0.1667 B
2	0.6667 A_2	0.3333 AB
3	0.5000 A_3	0.5000 A_2B
4	0.3333 A_4	0.6667 A_3B
5	0.1667 A_5	0.8333 A_4B

Formula: A_4B_2

1	0.6667 A	0.3333 B	0.0667 B_2
2	0.4000 A_2	0.5333 AB	0.2000 AB_2
3	0.2000 A_3	0.6000 A_2B	0.4000 A_2B_2
4	0.0667 A_4	0.5333 A_3B	0.6667 A_3B_2
5		0.3333 A_4B	

Formula: A_3B_3

1	0.5000 A	0.5000 B	0.2000 B_2	0.0500 B_3
2	0.2000 A_2	0.6000 AB	0.4500 AB_2	0.2000 AB_3
3	0.0500 A_3	0.4500 A_2B	0.6000 A_2B_2	0.5000 A_2B_3
4		0.2000 A_3B	0.5000 A_3B_2	
5				

Number of atoms scrambling : 7

Formula: A_6B

1	0.8571 A	0.1429 B
2	0.7143 A_2	0.2857 AB
3	0.5714 A_3	0.4286 A_2B
4	0.4286 A_4	0.5714 A_3B
5	0.2857 A_5	0.7143 A_4B
6	0.1429 A_6	0.8571 A_5B

Formula A_5B_2

1	0.7143 A	0.2857 B	0.0476 B_2
2	0.4762 A_2	0.4762 AB	0.1429 AB_2
3	0.2857 A_3	0.5714 A_2B	0.2857 A_2B_2
4	0.1429 A_4	0.5714 A_3B	0.4762 A_3B_2
5	0.0476 A_5	0.4762 A_4B	0.7143 A_4B_2
6		0.2857 A_5B	

Formula. A_4B_3

1	0.5714 A	0.4286 B	0.1429 B_2	0.0286 B_3
2	0.2857 A_2	0.5714 AB	0.3429 AB_2	0.1143 AB_3
3	0.1143 A_3	0.5143 A_2B	0.5143 A_2B_2	0.2857 A_2B_3
4	0.0286 A_4	0.3429 A_3B	0.5714 A_3B$_2$	0.5714 A_3B_3
5		0.1429 A_4B	0.4286 A_4B_2	

Number of atoms scrambling : 8
Formula: A_7B

1	0.8750 A	0.1250 B
2	0.7500 A_2	0.2500 AB
3	0.6250 A_3	0.3750 A_2B
4	0.5000 A_4	0.5000 A_3B
5	0.3750 A_5	0.6250 A_4B
6	0.2500 A_6	0.7500 A_5B
7	0.1250 A_7	0.8750 A_6B

Formula: A_6B_2

1	0.7500 A	0.2500 B	0.0357 B_2
2	0.5357 A_2	0.4286 AB	0.1071 AB_2
3	0.3571 A_3	0.5357 A_2B	0.2143 A_2B_2
4	0.2143 A_4	0.5714 A_3B	0.3571 A_3B_2
5	0.1071 A_5	0.5357 A_4B	0.5357 A_4B_2
6	0.0357 A_6	0.4286 A_5B	0.7500 A_5B_2
7		0.2500 A_6B	

Formula: A_5B_3

1	0.6250 A	0.3750 B	0.1071 B_2	0.0179 B_3
2	0.3571 A_2	0.5357 AB	0.2679 AB_2	0.0714 AB_3
3	0.1786 A_3	0.5357 A_2B	0.4286 A_2B_2	0.1786 A_2B_3
4	0.0714 A_4	0.4286 A_3B	0.5357 A_3B_2	0.3571 A_3B_3
5	0.0179 A_5	0.2679 A_4B	0.5357 A_4B_2	0.6250 A_4B_3
6		0.1071 A_5B	0.3750 A_5B_2	
7				

Formula: A_4B_4

1	0.5000 A	0.5000 B	0.2143 B_2	0.0714 B_3	0.0143 B_4
2	0.2143 A_2	0.5714 AB	0.4286 AB_2	0.2286 AB_3	0.0714 AB_4
3	0.0714 A_3	0.4286 A_2B	0.5143 A_2B_2	0.4286 A_2B_3	0.2143 A_2B_4
4	0.0143 A_4	0.2286 A_3B	0.4286 A_3B_2	0.5714 A_3B_3	0.5000 A_3B_4
5		0.0714 A_4B	0.2143 A_4B_2	0.5000 A_4B_3	
6					
7					

Number of atoms scrambling : 9

Formula: A_8B

1	0.8889 A	0.1111 B
2	0.7778 A_2	0.2222 AB
3	0.6667 A_3	0.3333 A_2B
4	0.5556 A_4	0.4444 A_3B
5	0.4444 A_5	0.5556 A_4B
6	0.3333 A_6	0.6667 A_5B
7	0.2222 A_7	0.7778 A_6B
8	0.1111 A_8	0.8889 A_7B

Formula: A_7B_2

1	0.7778 A	0.2222 B	0.0278 B_2
2	0.5833 A_2	0.3889 AB	0.0833 AB_2
3	0.4167 A_3	0.5000 A_2B	0.1667 A_2B_2
4	0.2778 A_4	0.5556 A_3B	0.2778 A_3B_2
5	0.1667 A_5	0.5556 A_4B	0.4167 A_4B_2
6	0.0833 A_6	0.5000 A_5B	0.5833 A_5B_2
7	0.0278 A_7	0.3889 A_6B	0.7778 A_6B_2
8		0.2222 A_7B	

Formula: A_6B_3

1	0.6667 A	0.3333 B	0.0833 B_2	0.0119 B_3
2	0.4167 A_2	0.5000 AB	0.2143 AB_2	0.0476 AB_3
3	0.2381 A_3	0.5357 A_2B	0.3571 A_2B_2	0.1190 A_2B_3
4	0.1190 A_4	0.4762 A_3B	0.4762 A_3B_2	0.2381 A_3B_3
5	0.0476 A_5	0.3571 A_4B	0.5357 A_4B_2	0.4167 A_4B_3
6	0.0119 A_6	0.2143 A_5B	0.5000 A_5B_2	0.6667 $A B$
7		0.0833 A_6B	0.3333 $A B$	
8				

Formula: A_5B_4

1	0.5556 A	0.4444 B	0.1667 B_2	0.0476 B_3	0.0079 B_4
2	0.2778 A_2	0.5556 AB	0.3571 AB_2	0.1587 AB_3	0.0397 AB_4
3	0.1190 A_3	0.4762 A_2B	0.4762 A_2B_2	0.3175 A_2B_3	0.1190 A_2B_4
4	0.0394 A_4	0.3175 A_3B	0.4762 A_3B_2	0.4762 A_3B_3	0.2778 A_3B_4
5	0.0049 A_5	0.1587 A_4B	0.3571 A_4B_2	0.5556 A_4B_3	0.5556 A_4B_4
6		0.0476 A_5B	0.1667 A_5B_2	0.4444 A_5B_3	
7					
8					

Number of atoms scrambling : 10

Formula: A_9B

1	0.9000 A	0.1000 B
2	0.8000 A_2	0.2000 AB
3	0.7000 A_3	0.3000 A_2B
4	0.6000 A_4	0.4000 A_3B
5	0.5000 A_5	0.5000 A_4B
6	0.4000 A_6	0.6000 A_5B
7	0.3000 A_7	0.7000 A_6B
8	0.2000 A_8	0.8000 A_7B
9	0.1000 A_9	0.9000 A_8B

Formula: A_8B_2

1	0.8000 A	0.2000 B	0.0222 B_2
2	0.6222 A_2	0.3556 AB	0.0667 AB_2
3	0.4667 A_3	0.4667 A_2B	0.1333 A_2B_2
4	0.3333 A_4	0.5333 A_3B	0.2222 A_3B_2
5	0.2222 A_5	0.5556 A_4B	0.3333 A_4B_2
6	0.1333 A_6	0.5333 A_5B	0.4667 A_5B_2
7	0.0667 A_7	0.4667 A_6B	0.6222 A_6B_2
8	0.0222 A_8	0.3556 A_7B	0.8000 A_7B_2
9		0.2000 A_8B	

Formula: A_7B_3

1	0.7000 A	0.3000 B	0.0667 B_2	0.0083 B_3
2	0.4667 A_2	0.4667 AB	0.1750 AB_2	0.0333 AB_3
3	0.2917 A_3	0.5250 A_2B	0.3000 A_2B_2	0.0833 A_2B_3
4	0.1667 A_4	0.5000 A_3B	0.4167 A_3B_2	0.1667 A_3B_3
5	0.0833 A_5	0.4167 A_4B	0.5000 A_4B_2	0.2917 A_4B_3
6	0.0333 A_6	0.3000 A_5B	0.5250 A_5B_2	0.4667 A_5B_3
7	0.0083 A_7	0.1750 A_6B	0.4667 A_6B_2	0.7000 A_6B_3
8		0.0667 A_7B	0.3000 A_7B_2	
9				

Formula: A_6B_4

1	0.6000 A	0.4000 B	0.1333 B_2	0.0333 B_3	0.0048 B_4
2	0.3333 A_2	0.5333 AB	0.3000 AB_2	0.1143 AB_3	0.0238 AB_4
3	0.1667 A_3	0.5000 A_2B	0.4286 A_2B_2	0.2381 A_2B_3	0.0714 A_2B_4
4	0.0714 A_4	0.3810 A_3B	0.4762 A_3B_2	0.3810 A_3B_3	0.1667 A_3B_4
5	0.0238 A_5	0.2381 A_4B	0.4286 A_4B_2	0.5000 A_4B_3	0.3333 A_4B_4
6	0.0048 A_6	0.1143 A_5B	0.3000 A_5B_2	0.5333 A_5B_3	0.6000 A_5B_4
7		0.0333 A_6B	0.1333 A_6B_2	0.4000 A_6B_3	
8					
9					

Formula: A_5B_5

1	0.5000 A	0.5000 B	0.2222 B_2	0.0833 B_3	0.0238 B_4	0.0040 B_5
2	0.2222 A_2	0.5556 AB	0.4167 AB_2	0.2381 AB_3	0.0992 AB_4	0.0238 AB_5
3	0.0833 A_3	0.4167 A_2B	0.4762 A_2B_2	0.3968 A_2B_3	0.2381 A_2B_4	0.0833 A_2B_5
4	0.0238 A_4	0.2381 A_3B	0.3968 A_3B_2	0.4762 A_3B_3	0.4167 A_3B_4	0.2222 A_3B_5
5	0.0040 A_5	0.0992 A_4B	0.2381 A_4B_2	0.4167 A_4B_3	0.5556 A_4B_4	0.5000 A_4B_5
6		0.0238 A_5B	0.0833 A_5B_2	0.2222 A_5B_3	0.5000 A_5B_4	

Pressure and energy units

Pressure
    1 Pascal (Pa) = 1 Newton.m^{-2} = 7.501 x 10^{-3} mm Hg
    1 torr = 133.3 Pa
    1 torr = 1 mm Hg (accurate to 1 part in 10^7)
    1 barye = 1 dyne.cm^{-2} = 7.501 x 10^{-4} mm Hg
    1 micron = 10^{-3} mm Hg

Energy
    1 eV = 23.061 kcal.mole^{-1}
    1 eV = 1.602 x 10^{-12} erg
    1 eV = 96.487 kJ.mole^{-1}

Commonly used metastable ion formulae

These formulae are collected here for convenience. In all cases a metastable ion, mass m_1, carrying x charges, fragments to give a fragment ion mass m_2 carrying y charges and a neutral fragment, mass m_3. The electric sector voltage is E, the ion accelerating voltage, V.

Position of a metastable peak in the mass spectrum: $\dfrac{xm_2{}^2}{y^2 m_1}$

Position of the peak in the IKE spectrum: $\dfrac{m_2 xE}{m_1 y}$

Position of the peak in the HV scan: $\dfrac{m_1 yV}{m_2 x}$

Relationship between kinetic energy release and metastable peak width, d, in the mass spectrum

$$T = \frac{y^4 m_1{}^2 d^2 eV}{16 x m_2{}^3 m_3}$$

Relationship between kinetic energy release and peak width, ΔE, in an IKE or MIKE spectrum

$$T = \frac{y^2 m_1{}^2 eV}{16 x m_2 m_3} \left(\frac{\Delta E}{E}\right)^2$$

Relationship between kinetic energy release and peak width, ΔV, in an HV scan spectrum

$$T = \frac{x^2 m_2{}^2 eV}{16 y m_1 m_3} \left(\frac{\Delta V}{V}\right)^2$$

Amplification factor, A, operating on energy spread of the products of a metastable ion fragmentation

$$A = 4 \left(\frac{xeV}{T}\right)^{\frac{1}{2}} \frac{(m_2 m_3)^{\frac{1}{2}}}{m_1}$$

Relationship between kinetic energy release (eV) and intercharge distance (Å) in a doubly charged ion

$$T = \frac{14.39}{R}$$

References

"All I know is what I read in the papers"

Will Rogers

1   A.J. Ahearn (Ed.), *Mass Spectrometric Analysis of Solids*, Elsevier, Amsterdam, 1966.
2   M.I. Al-Joboury and D.W. Turner, *J. Chem. Soc. (London)*, (1964) 4434.
3   S.L. Altmann, *Proc. Roy. Soc. (London), Ser. A*, 298 (1967) 184.
4   B. Andlauer and Ch. Ottinger, *J. Chem. Phys.*, 55 (1971) 1471.
5   B. Andlauer and Ch. Ottinger, *Z. Naturforsch. A*, 27 (1972) 293.
6   J. Appell, P.G. Fournier, F.C. Fehsenfeld and J. Durup, *20th Annual Conference on Mass Spectrometry and Allied Topics, Dallas, 1972*, American Society for Mass Spectrometry.
7   T. Ast, *Ph.D. Thesis*, Purdue University, 1972.
8   T. Ast, J.H. Beynon and R.G. Cooks, *J. Amer. Chem. Soc.*, 94 (1972) 1834.
9   T. Ast, J.H. Beynon and R.G. Cooks, *Org. Mass Spectrom.*, 6 (1972) 741.
10 T. Ast, J.H. Beynon and R.G. Cooks, *Org. Mass Spectrom.*, 6 (1972) 749.
11 T. Ast and J.H. Beynon, *Org. Mass Spectrom.*, 7 (1973) 503.
12 T. Ast, J.H. Beynon and R.G. Cooks, *J. Amer. Chem. Soc.*, 94 (1972) 6611.
13 J.D. Baldeschwieler, *Science*, 159 (1968) 263.
14 J.D. Baldeschwieler and S.S. Woodgate, *Accounts Chem. Res.*, 4 (1971) 114.
15 M. Barber and R.M. Elliott, *12th Annual Conference on Mass Spectrometry and Allied Topics, Montreal, 1964*, ASTM Committee E-14.
16 M. Barber, W.A. Wolstenholme and K.R. Jennings, *Nature*, 214 (1967) 664.
17 N.F. Barber, *Proc. Leeds Phil. Lit. Soc., Sci. Sect.*, 2 (1933) 427.
18 C.S. Barnes and J.L. Occolowitz, *Aust. J. Chem.*, 16 (1963) 219.
19 J.L. Beauchamp and S.E. Buttrill, *J. Chem. Phys.*, 48 (1968) 1783.
20 J.L. Beauchamp and R.C. Dunbar, *J. Amer. Chem. Soc.*, 92 (1970) 1477.
21 H.D. Beckey, *Advan. Mass Spectrom.*, 2 (1963) 1.
22 H.D. Beckey, *Z. Naturforsch. A*, 19 (1964) 71.
23 H.D. Beckey, H. Hey, K. Levsen and G. Tenschert, *Int. J. Mass Spectrom. Ion Phys.*, 2 (1969) 101.
24 R.P. Bell, *Trans. Faraday Soc.*, 55 (1959) 1.
25 D.A. Ben-Efraim, C. Batich and E. Wasserman, *J. Amer. Chem. Soc.*, 92 (1970) 2133.
26 S.A. Benezra, M.K. Hoffman and M.M. Bursey, *J. Amer. Chem. Soc.*, 92 (1970) 7501.
27 S.A. Benezra and M.M. Bursey, *J. Amer. Chem. Soc.*, 94 (1972) 1024.
28 H. Benz and H.W. Brown, *J. Chem. Phys.*, 48 (1968) 4308.
29 W. Benz and K. Biemann, *J. Amer. Chem. Soc.*, 86 (1964) 2375.
30 M. Bertrand, J.H. Beynon and R.G. Cooks, *Int. J. Mass Spectrom. Ion Phys.*, 9 (1972) 346.
31 M. Bertrand, J.H. Beynon and R.G. Cooks, *Org. Mass Spectrom.*, 7 (1973) 193.
32 J.H. Beynon, B.E. Job and A.E. Williams, *Z. Naturforsch. A*, 20 (1965) 883.
33 J.H. Beynon, *Mass Spectrometry and Its Applications to Organic Chemistry*, Elsevier, Amsterdam, 1960.

34 J.H. Beynon, G.R. Lester, R.A. Saunders and A.E. Williams, *Trans. Faraday Soc.*, 57 (1961) 1259.

35 J.H. Beynon, R.A. Saunders and A.E. Williams, *Table of Meta-Stable Transitions for Use in Mass Spectrometry*, Elsevier, Amsterdam, 1965.

36 J.H. Beynon, R.A. Saunders and A.E. Williams, *Z. Naturforsch. A*, 20 (1965) 180.

37 J.H. Beynon and A.E. Fontaine, *Chem. Commun.*, (1966) 717.

38 J.H. Beynon and A.E. Fontaine, *Z. Naturforsch. A*, 22 (1967) 334.

39 J.H. Beynon, A.E. Fontaine and G.R. Lester, *Int. J. Mass Spectrom. Ion Phys.*, 1 (1968) 1.

40 J.H. Beynon, J.A.V. Hopkinson and G.R. Lester, *Int. J. Mass Spectrom. Ion Phys.*, 1 (1968) 343.

41 J.H. Beynon, *Advan. Mass Spectrom.*, 4 (1968) 123.

42 J.H. Beynon, R.A. Saunders and A.E. Williams, *The Mass Spectra of Organic Molecules*, Elsevier, New York, 1968.

43 J.H. Beynon, J.A. Hopkinson and A.E. Williams, *Org. Mass Spectrom.*, 1 (1968) 169.

44 J.H. Beynon, W.E. Baitinger, J.W. Amy and T. Komatsu, *Int. J. Mass Spectrom. Ion Phys.*, 3 (1969) 47.

45 J.H. Beynon, J.W. Amy and W.E. Baitinger, *Int. J. Mass Spectrom. Ion Phys.*, 3 (1969) 55.

46 J.H. Beynon, J.A. Hopkinson and G.R. Lester, *Int. J. Mass Spectrom. Ion Phys.*, 2 (1969) 291.

47 J.H. Beynon, R.M. Caprioli, W.E. Baitinger and J.W. Amy, *Int. J. Mass Spectrom. Ion Phys.*, 3 (1969) 313.

48 J.H. Beynon, J.W. Amy and W.E. Baitinger, *Chem. Commun.*, (1969) 723.

49 J.H. Beynon, R.M. Caprioli, W.E. Baitinger and J.W. Amy, *Org. Mass Spectrom.*, 3 (1970) 455.

50 J.H. Beynon, R.M. Caprioli, W.E. Baitinger and J.W. Amy, *Org. Mass Spectrom.*, 3 (1970) 479.

51 J.H. Beynon, R.M. Caprioli, W.E. Baitinger and J.W. Amy, *Org. Mass Spectrom.*, 3 (1970) 661.

52 J.H. Beynon, J.E. Corn, W.E. Baitinger, J.W. Amy and R.A. Benkeser, *Org. Mass Spectrom.*, 3 (1970) 1371.

53 J.H. Beynon, R.M. Caprioli and T. Ast, *Int. J. Mass Spectrom. Ion Phys.*, 7 (1971) 88.

54 J.H. Beynon, R.M. Caprioli and T. Ast, *Org. Mass Spectrom.*, 5 (1971) 229.

55 J.H. Beynon, A. Mathias and A.E. Williams, *Org. Mass Spectrom.*, 5 (1971) 303.

56 J.H. Beynon and R.G. Cooks, *Res./Develop.*, 22 (11) (1971) 26.

57 J.H. Beynon, M. Bertrand, E.G. Jones and R.G. Cooks, *Chem. Commun.*, (1972) 341.

58 J.H. Beynon, A.E. Fontaine and G.R. Lester, *Int. J. Mass Spectrom. Ion Phys.*, 8 (1972) 341.

59 J.H. Beynon, R.M. Caprioli and T. Ast, *Org. Mass Spectrom.*, 6 (1972) 273.

60 J.H. Beynon, R.M. Caprioli, W.O. Perry and W.E. Baitinger, *J. Amer. Chem. Soc.*, 94 (1972) 6828.

61 J.H. Beynon, R.M. Caprioli, R.H. Shapiro, K.B. Tomer and C.W. Chang, *Org. Mass Spectrom.*, 6 (1972) 863.

62 J. H. Beynon, M. Bertrand and R. G. Cooks, *J. Amer. Chem. Soc.*, 95 (1973) 1739.

63 J.H. Beynon, R.G. Cooks and M. Bertrand, *Org. Mass. Spectrom.*, in press.
64 J.H. Beynon and R.G. Cooks, in preparation.
65 K. Biemann and J. Seibl, *J. Amer. Chem. Soc.*, 81 (1959) 3149.
66 K. Biemann, *Mass Spectrometry: Organic Chemical Applications*, McGraw-Hill, New York, 1962.
67 E.W. Blauth, *Dynamic Mass Spectrometers*, Elsevier, Amsterdam, 1966.
68 M. Born and R. Oppenheimer, *Ann. Physik.*, 84 (1927) 457.
69 P.R. Briggs, W.L. Parker and T.W. Shannon, *Chem. Commun.*, (1968) 727.
70 C.E. Brion and L.D. Hall, *J. Amer. Chem. Soc.*, 88 (1966) 3661.
71 P. Brown and C. Djerassi, *J. Amer. Chem. Soc.*, 89 (1967) 2711.
72 P. Brown, *Org. Mass Spectrom.*, 4 (1970) 519.
73 H. Budzikiewicz, C. Djerassi and D.H. Williams, *Interpretation of Mass Spectra of Organic Compounds*, Holden-Day, San Francisco, 1964.
74 H. Budzikiewicz, C. Djerassi and D.H. Williams, *Mass Spectrometry of Organic Compounds*, Holden-Day, San Francisco, 1967.
75 M.M. Bursey and F.W. McLafferty, *J. Amer. Chem. Soc.*, 88 (1966) 5023.
76 M.M. Bursey and F.W. McLafferty, in G.A. Olah and P. von R. Schleyer (Eds.), *Carbonium Ions*, Vol. 1, Interscience; New York, 1968, p. 257.
77 M.M. Bursey, F.E. Tibbetts, W.F. Little, M.D. Rausch and G.A. Moser, *Tetrahedron Lett.*, (1969) 3469.
78 M.M. Bursey, M.K. Hoffman and S.A. Benezra, *Chem. Commun.*, (1971) 1417.
79 R.M. Caprioli, J.H. Beynon and T. Ast, *Org. Mass Spectrom.*, 5 (1971) 417.
80 E.M. Chait and W.B. Askew, *Org. Mass Spectrom.*, 5 (1971) 147.
81 J.R. Chapman, S. Evans and W.A. Wolstenholme, *18th Annual Conference on Mass Spectrometry and Allied Topics, San Francisco, 1970*, American Society for Mass Spectrometry.
82 W.A. Chupka, *J. Chem. Phys.*, 30 (1959) 191.
83 W.A. Chupka, J. Berkowitz and S.I. Miller, *20th Annual Conference on Mass Spectrometry and Allied Topics, Dallas, 1972*, American Society for Mass Spectrometry.
84 N.D. Coggeshall, *J. Chem. Phys.*, 37 (1962) 2167.
85 R.G. Cooks, I. Howe and D.H. Williams, *Org. Mass Spectrom.*, 2 (1969) 137.
86 R.G. Cooks and S.L. Bernasek, *J. Amer. Chem. Soc.*, 92 (1970) 2129.
87 R.G. Cooks and J.H. Beynon, *Chem. Commun.*, (1971) 1282.
88 R.G. Cooks, J.H. Beynon and T. Ast, *J. Amer. Chem. Soc.*, 94 (1972) 1004.
89 R.G. Cooks, M. Bertrand, J.H. Beynon, M.E. Rennekamp and D.W. Setser, *J. Amer. Chem. Soc.*, 95 (1973) 1732.
90 R.G. Cooks, T. Ast and J.H. Beynon, *Int. J. Mass Spectrom., Ion Phys.*, in press.
91 R.G. Cooks, M. Bertrand, J.H. Beynon and M.K. Hoffman, submitted for publication.
92 C.A. Coulson and K. Zalewski, *Proc. Roy. Soc. (London), Ser. A*, 268 (1962) 437.
93 R.D. Craig, B.N. Green and J.D. Waldron, *Chimia*, 17 (1963) 33.
94 N.R. Daly, *Proc. Phys. Soc. (London)*, 85 (1965) 897.
95 B. Davis, D.H. Williams and A.N.H. Yeo, *J. Chem. Soc. B*, (1970) 81.
96 F. DeJong, H.J.M. Sinnige and M.J. Janssen, *Org. Mass Spectrom.*, 3 (1970) 1539.
97 J. Delwiche and P. Natalis, *Chem. Phys. Lett.*, 5 (1970) 564.

98 P.J. Derrick, A.M. Falick and A.L. Burlingame, *20th Annual Conference on Mass Spectrometry and Allied Topics, Dallas, 1972*, American Society for Mass Spectrometry.

99 R.J. Dickerson and D.H. Williams, *J. Chem. Soc. B*, (1971) 249.

100 J. Diekman, J.K. MacLeod, C. Djerassi and J.D. Baldeschwieler, *J. Amer. Chem. Soc.*, 91 (1969) 2069.

101 J. Durup, P. Fournier and P. Dông, *Int. J. Mass Spectrom. Ion Phys.*, 2 (1969) 311.

102 J. Durup, in K. Ogata and T. Hayakawa (Eds.), *Recent Developments in Mass Spectroscopy*, University Park Press, Baltimore, 1970, p. 921.

103 H.B. Dwight, *Tables of Integrals and Other Mathematical Data*, Macmillan, New York 1957, p. 7.

104 G. Eadon, J. Diekman and C. Djerassi, *J. Amer. Chem. Soc.*, 91 (1969) 3986.

105 G. Eadon, C. Djerassi, J.H. Beynon and R.M. Caprioli, *Org. Mass Spectrom.*, 5 (1971) 917.

106 H. Ewald and W. Seibt, in K. Ogata and T. Hayakawa (Eds.), *Recent Developments in Mass Spectroscopy*, University Park Press, Baltimore, 1970, p. 39.

107 C.C. Fenselau and C.H. Robinson, *J. Amer. Chem. Soc.*, 93 (1971) 3070.

108 F.H. Field and M.S.B. Munson, *J. Amer. Chem. Soc.*, 89 (1967) 1047.

109 F.H. Field, *Accounts Chem. Res.*, 1 (1968) 42.

110 F. Fiquet-Fayard and P.M. Guyon, *Mol. Phys.*, 11 (1966) 17.

111 F. Fiquet-Fayard, *Israel J. Chem.*, 7 (1969) 275.

112 M.C. Flowers, *Chem. Commun.*, (1965) 235.

113 J.L. Franklin, in G.A. Olah and P. von R. Schleyer (Eds.), *Carbonium Ions*, Vol. 1, Interscience, New York, 1968, p. 77.

114 J.L. Franklin, J.G. Dillard, H.M. Rosenstock, J.T. Herron, K. Draxl and F.H. Field, Ionization potentials, appearance potentials and heats of formation of gaseous positive ions, *NSRDS-NBS No. 26*, National Bureau of Standards, Washington, D.C., 1969.

115 J.L. Franklin and M.A. Haney, in K. Ogata and T. Hayakawa (Eds.), *Recent Developments in Mass Spectroscopy*, University Park Press, Baltimore, 1970, p. 909.

116 D.C. Frost, A. Katrib, C.A. McDowell and R.A.N. McLean, *Int. J. Mass Spectrom. Ion Phys.*, 7 (1971) 485.

117 J.H. Futrell, K.R. Ryan and L.W. Sieck, *J. Chem. Phys.*, 43 (1965) 1832.

118 E. Gil-Av, J.H. Leftin, A. Mandelbaum and S. Weinstein, *Org. Mass Spectrom.*, 4 (1970) 475.

119 T.R. Govers and J. Schopman, *Chem. Phys. Lett.*, 12 (1971) 414.

120 M.M. Green, R.J. Cook, J.M. Schwab and R.B. Roy, *J. Amer. Chem. Soc.*, 92 (1970) 3076.

121 M.M. Green, J.M. Moldowan, D.J. Hart and J.M. Krakower, *J. Amer. Chem. Soc.*, 92 (1970) 3491.

122 M.M. Green, J.G. McGrew and J.M. Moldowan, *J. Amer. Chem. Soc.*, 93 (1971) 6700.

123 M.L. Gross and F.W. McLafferty, *J. Amer. Chem. Soc.*, 93 (1971) 1267.

124 W.F. Haddon and F.W. McLafferty, *J. Amer. Chem. Soc.*, 90 (1968) 4745.

125 W.F. Haddon and F.W. McLafferty, *Anal. Chem.*, 41 (1969) 31.

126 M.A. Haney and J.L. Franklin, *J. Chem. Phys.*, 48 (1968) 4093.

127 M.A. Haney and J.L. Franklin, *Trans. Faraday Soc.*, 65 (1969) 1794.
128 M.A. Haney and R.T. McIver, *19th Annual Conference on Mass Spectrometry and Allied Topics, Atlanta, 1971,* American Society for Mass Spectrometry.
129 A.G. Harrison, in A.L. Burlingame (Ed.), *Topics in Organic Mass Spectrometry*, Wiley-Interscience, New York, 1970, p. 121.
130 A.G. Harrison, C.D. Finney and J.A. Sherk, *Org. Mass Spectrom.*, 5 (1971) 1313.
131 J.B. Hasted, *Physics of Atomic Collisions*, Butterworths, London, 1964, p. 3.
132 I. Hertel and Ch. Ottinger, *Z. Naturforsch. A*, 22 (1967) 40.
133 G. Herzberg, *Infra-Red and Raman Spectra of Polyatomic Molecules*, D. van Nostrand, New York, 1945, p. 215.
134 R.F.K. Herzog, *Z. Physik*, 89 (1934) 447.
135 R.D. Hickling and K.R. Jennings, *Org. Mass Spectrom.*, 3 (1970) 1499.
136 W. Higgins and K.R. Jennings, *Chem. Commun.*, (1965) 99.
137 W. Higgins and K.R. Jennings, *Trans. Faraday Soc.*, 62 (1966) 97.
138 H.C. Hill, *Introduction to Mass Spectrometry*, Heyden and Son, London, 1966.
139 H. Hintenberger (Ed.), *Nuclear Masses and Their Determination*, Pergamon Press, London, 1957.
140 J.A. Hipple and E.U. Condon, *Phys. Rev.*, 68 (1945) 54.
141 J.A. Hipple, R.E. Fox and E.U. Condon, *Phys. Rev.*, 69 (1946) 347.
142 M.K. Hoffman and M.M. Bursey, *Tetrahedron Lett.*, (1971) 2539.
143 I. Horman, A.N.H. Yeo and D.H. Williams, *J. Amer. Chem. Soc.*, 92 (1970) 2131.
144 J.C. Houver, J. Baudon, M. Abignoli, M. Barat, P. Fournier and J. Durup, *Int. J. Mass Spectrom. Ion Phys.*, 4 (1970) 137.
145 I. Howe and D.H. Williams, *J. Amer. Chem. Soc.*, 91 (1969) 7137.
146 I. Howe, in D.H. Williams (Ed.), *Mass Spectrometry*, Vol. 1, *A Specialist Periodical Report*, The Chemical Society, London, 1971 p. 31.
147 I. Howe and F.W. McLafferty, *J. Amer. Chem. Soc.*, 93 (1971) 99.
148 I. Howe and D.H. Williams, *Chem. Commun.*, (1971) 1195.
149 M.G. Inghram and R. Gomer, *J. Chem. Phys.*, 22 (1954) 1279.
150 M.G. Inghram and R. Gomer, *Z. Naturforsch. A*, 10 (1955) 863.
151 R. Jayaram, *Mass Spectrometry, Theory and Applications*, Plenum Press, New York, 1966.
152 K.R. Jennings, *J. Chem. Phys.*, 43 (1965) 4176.
153 K.R. Jennings, *Z. Naturforsch. A*, 22 (1967) 454.
154 K.R. Jennings, *Int. J. Mass Spectrom. Ion Phys.*, 1 (1968) 227.
155 E.G. Jones, L.E. Bauman, R.G. Cooks and J.H. Beynon, *Org. Mass Spectrom.*, 7 (1973) 185.
156 E.G. Jones, J.H. Beynon and R.G. Cooks, *J. Chem. Phys.*, 57 (1972) 2652.
157 E.G. Jones, J.H. Beynon and R.G. Cooks, *J. Chem. Phys.*, 57 (1972) 3207.
158 L.S. Kassel, *J. Phys. Chem.*, 32 (1928) 225.
159 T. Keough, T. Ast, J.H. Beynon and R.G. Cooks, *Org. Mass Spectrom.*, 7 (1973) 245.
160 T. Keough, J.H. Beynon and R.G. Cooks, *J. Amer. Chem. Soc.*, 95 (1973) 1695
161 R.W. Kiser, *Introduction to Mass Spectrometry and Its Applications*, Prentice-Hall, Englewood Cliffs, N.J., 1965.

162   R.W. Kiser, R.E. Sullivan and M.S. Lupin, *Anal. Chem.*, 41 (1969) 1958.
163   C.E. Klotz, *J. Chem. Phys.*, 41 (1964) 117.
164   A Kropf, E.M. Eyring, A.L. Wahrhaftig and H. Eyring, *J. Chem. Phys.*, 32 (1960) 149.
165   M.J. Lacey, C.G. MacDonald and J.S. Shannon, *Org. Mass Spectrom.*, 5 (1971) 1391.
166   K.J. Laidler, *The Chemical Kinetics of Excited States*, Oxford University Press, Oxford, 1955.
167   L. Landau, *Phys. Z. Sowjetunion*, 2 (1932) 88.
168   G.F. Lanthier and J.M. Miller, *Org. Mass Spectrom.*, 6 (1972) 89.
169   B.S. Larsen, G. Schroll, S.-O. Lawesson, J.H. Bowie and R.G. Cooks, *Tetrahedron*, 24 (1968) 5193.
170   C. Lifshitz and R. Sternberg, *Int. J. Mass Spectrom. Ion Phys.*, 2 (1969) 303.
171   Y.N. Lin, and B.S. Rabinovitch, *J. Phys. Chem.*, 74 (1970) 1769.
172   H.C. Longuet-Higgins, *Mol. Phys.*, 6 (1963) 445.
173   H.W. Major, *Docket No. D-1185, A626*, U.S. Patent Office, Washington, D.C., May 1st, 1968.
174   M.M. Mann, A Hustrulid and J.T. Tate, *Phys. Rev.*, 58 (1940) 340.
175   N.H. March and A.M. Murray, *Proc. Roy. Soc. (London), Ser. A*, 261 (1961) 119.
176   R.A. Marcus, *J. Chem. Phys.*, 20 (1952) 359.
177   H.S.W. Massey and E.H.S. Burhop, *Electronic and Ionic Impact Phenomena*, Oxford University Press, Oxford, 1952, p. 533.
178   J. Mattauch and R.F.K. Herzog, *Z. Physik*, 89 (1934) 786.
179   K.H. Maurer, C. Brunnee, G. Kappus, K. Habfast, U. Schroder and P. Schulze, *19th Annual Conference on Mass Spectrometry and Allied Topics, Atlanta, 1971*, American Society for Mass Spectrometry.
180   D.J. McAdoo, F.W. McLafferty and P.F. Bente, *J. Amer. Chem. Soc.*, 94 (1972) 2027.
181   C.G. McDonald and J.S. Shannon, *Aust. J. Chem.*, 15 (1962) 771.
182   C.A. McDowell, in *Applied Mass Spectrometry: Proceedings of a Mass Spectrometry Conference*, Institute of Petroleum, London, 1953, p. 129.
183   C.A. McDowell, (Ed.), *Mass Spectrometry*, McGraw-Hill, New York, 1963.
184   J.W. McGowan and L. Kerwin, *Can. J. Phys.*, 41 (1963) 1535.
185   F.W. McLafferty, *Anal. Chem.*, 31 (1959) 82.
186   F.W. McLafferty, in F.W. McLafferty (Ed.), *Mass Spectra of Organic Ions*, Academic Press, New York, 1963, p. 309.
187   F.W. McLafferty, *Interpretation of Mass Spectra*, Benjamin, New York, 1966.
188   F.W. McLafferty, M.M. Bursey and S.M. Kimball, *J. Amer. Chem. Soc.*, 88 (1966) 5022.
189   F.W. McLafferty and T.A. Bryce, *Chem. Commun.*, (1967) 1215.
190   F.W. McLafferty and W.T. Pike, *J. Amer. Chem. Soc.*, 89 (1967) 5951.
191   F.W. McLafferty and W.T. Pike, *J. Amer. Chem. Soc.*, 89 (1967) 5953.
192   F.W. McLafferty and M.M. Bursey, *J. Org. Chem.*, 33 (1968) 124.
193   F.W. McLafferty and R.B. Fairweather, *J. Amer. Chem. Soc.*, 90 (1968) 5915.
194   F.W. McLafferty and H.D.R. Schuddemage, *J. Amer. Chem. Soc.*, 91 (1969) 1866.
195   F.W. McLafferty, R. Venkataraghavan and P. Irving, *Biochem. Biophys. Res. Commun.*, 39 (1970) 274.

196 F.W. McLafferty, in A.L. Burlingame (Ed.), *Topics in Organic Mass Spectrometry*, Wiley-Interscience, New York, 1970, p. 223.

197 F.W. McLafferty, T. Wachs, C. Lifshitz, G. Innorta and P. Irving, *J. Amer. Chem. Soc.*, 92 (1970) 6867.

198 F.W. McLafferty, D.J. McAdoo, J.S. Smith and R. Kornfeld, *J. Amer. Chem. Soc.*, 92 (1971) 3720.

199 **F.W. McLafferty, P.F. Bente, III, R. Kornfeld, S.-C. Tsai and I. Howe, *J. Amer. Chem. Soc.*, 95 (1973) 2120.**

200 C.E. Melton, in F.W. McLafferty (Ed.), *Mass Spectrometry of Organic Ions*, Academic Press, New York, 1963, p. 163.

201 F. Meyer and A.G. Harrison, *J. Amer. Chem. Soc.*, 86 (1964) 4757.

202 S. Meyerson, P.N. Rylander, E.L. Eliel and J.D. McCollum, *J. Amer. Chem. Soc.*, 81 (1959) 2606.

203 S. Meyerson and J.D. McCollum, *Advan. Anal. Chem. Instrum.*, 2 (1963) 179.

204 S. Meyerson, *J. Amer. Chem. Soc.*, 85 (1963) 3340.

205 S. Meyerson and L.C. Leitch, *J. Amer. Chem. Soc.*, 86 (1964) 2555.

206 S. Meyerson and A.W. Weitkamp, *Org. Mass Spectrom.*, 1 (1968) 659.

207 S. Meyerson and L.C. Leitch, *J. Amer. Chem. Soc.*, 93, (1971) 2244.

208 J. Momigny, L. Brakier and L. D'Or, *Bull. Cl. Sci. Acad. Roy. Belg.*, 48 (1962) 1002.

209 T.F. Moran, F.C. Petty and A.F. Hedrick, *J. Chem. Phys.*, 51 (1969) 2112.

210 N.F. Mott and I.N. Sneddon, *Wave Mechanics and Its Applications*, Oxford University Press, Oxford, 1948, p. 17.

211 G.A. Muccini, W.H. Hamill and R. Barker, *J. Phys. Chem.*, 68 (1964) 261.

212 M.E. Munk, C.L. Kulkarni, C.L. Lee and P. Brown, *Tetrahedron*, (1970) 1377.

213 M.S.B. Munson and F.H. Field, *J. Amer. Chem. Soc.*, 88 (1966) 2621.

214 A.S. Newton and A.F. Sciamanna, *J. Chem. Phys.*, 40 (1964) 718.

215 A.S. Newton, *J. Chem. Phys.*, 44 (1966) 4015.

216 A.S. Newton, A.F. Sciamanna and R. Clampitt, *J. Chem. Phys.*, 46 (1967) 1779.

217 A.S. Newton, A.F. Sciamanna and R. Clampitt, *J. Chem. Phys.*, 47 (1967) 4843.

218 A.S. Newton and A.F. Sciamanna, *J. Chem. Phys.*, 50 (1969) 4868.

219 A.O.C. Nier, *Rev. Sci. Instrum.*, 11 (1940) 212.

220 Ch. Ottinger, *Z. Naturforsch. A*, 22 (1967) 20.

221 J.L. Occolowitz, *J. Amer. Chem. Soc.*, 91 (1969) 5202.

222 J.E. Parker and R.S. Lehrle, *Int. J. Mass Spectrom. Ion Phys.*, 7 (1971) 421.

223 W.O. Perry, J.H. Beynon, W.E. Baitinger, J.W. Amy, R.M. Caprioli, R.N. Renaud, L.C. Leitch and S. Meyerson, *J. Amer. Chem. Soc.*, 92 (1970) 7236.

224 A.W. Potts and W.C. Price, *Proc. Roy. Soc. (London), Ser. A*, 326 (1972) 181.

225 B.S. Rabinovitch and D.W. Setser, *Advan. Photochem.*, 3 (1964) 1.

226 R.I. Reed, *Application of Mass Spectrometry to Organic Chemistry*, Academic Press, New York, 1966.

227 M.E. Rennekamp, W.O. Perry and R.G. Cooks, *J. Amer. Chem. Soc.*, 94 (1972) 4985.

228 O.K. Rice and H.C. Ramsperger, *J. Amer. Chem. Soc.*, 49 (1927) 1617.
229 J. Roboz, *Introduction to Mass Spectrometry. Instrumentation and Techniques*, Interscience, New York, 1968.
230 F.W. Röllgen and H.D. Beckey, *Surface Sci.*, 23 (1970) 69.
231 H.M. Rosenstock, M.B. Wallenstein, A.L. Wahrhaftig and H. Eyring, *Proc. Nat. Acad. Sci. U.S.*, 38 (1952) 667.
232 H.M. Rosenstock and C.E. Melton, *J. Chem. Phys.*, 26 (1957) 314.
233 H.M. Rosenstock and M. Krauss, in F.W. McLafferty (Ed.), *Mass Spectra of Organic Ions*, Academic Press, New York, 1963, p. 1.
234 H.M. Rosenstock, V.H. Dibeler and F.N. Harlee, *J. Chem. Phys.*, 40 (1964) 591.
235 H.M. Rosenstock, *Advan. Mass Spectrom.*, 4 (1968) 523.
236 F.M. Rourke, J.C. Sheffield, W.D. Davis and F.A. White, *J. Chem. Phys.*, 31 (1959) 193.
237 C.G. Rowland, J.H.D. Eland and C.J. Danby, *Chem. Commun.*, (1968) 1535.
238 S. Safe, O. Huntzinger and W.D. Jamieson, *Org. Mass Spectrom.*, 7 (1973) 169.
239 J. Schopman, A.K. Barua and J. Los, in K. Ogata and T. Hayakawa (Eds.), *Recent Developments in Mass Spectroscopy*, University Park Press, Baltimore, 1970, p. 888.
240 J. Seibl, *Org. Mass Spectrom.*, 2 (1969) 1033.
241 J.S. Shannon, *Proc. Roy. Aust. Chem. Inst.*, (1964) 328.
242 T.W. Shannon and F.W. McLafferty, *J. Amer. Chem. Soc.*, 88 (1966) 5021.
243 T.W. Shannon, *Int. J. Mass Spectrom. Ion Phys.*, 3 (1970) App.12.
244 S.R. Shrader, *Introductory Mass Spectrometry*, Allyn and Bacon, Boston, 1971.
245 A.S. Siegel, *Tetrahedron Lett.*, (1970) 4113.
246 N.B. Slater, *Theory of Unimolecular Reactions*, Methuen, London, 1959.
247 G.A. Smith and D.H. Williams, *J. Chem. Soc. B*, (1970) 1529.
248 G. Spiteller, *Massenspektrometrische Strukturanalyse Organischer Verbindungen*, Verlag Chemie, Weinheim, 1966.
249 B. Steiner, C.F. Giese and M.G. Inghram, *J. Chem. Phys.*, 34 (1961) 189.
250 W.E. Stephens and A.L. Hughes, *Phys. Rev.*, 45 (1934) 123.
251 W.E. Stephens, *Phys. Rev.*, 45 (1934) 513.
252 D.P. Stevenson, *Discuss. Faraday Soc.*, 10 (1951) 35.
253 T. Sun and R.E. Lovins, *Org. Mass Spectrom.*, 6 (1972) 39.
254 H.J. Svec and G.D. Flesch, *Int. J. Mass Spectrom. Ion Phys.*, 1 (1968) 41.
255 R. Taubert, *Z. Naturforsch. A*, 19 (1964) 911.
256 T.O. Tiernan and C. Lifshitz, *Advan. Mass Spectrom.*, 5 (1971) 184.
257 T.O. Tiernan and L.P. Hills, *19th Annual Conference on Mass Spectrometry and Allied Topics, Atlanta, 1971*, American Society for Mass Spectrometry.
258 R.C. Tolman, *The Principles of Statistical Mechanics*, Oxford University Press, Oxford, 1938, p. 55.
259 J.C. Tou, *J. Phys. Chem.*, 74 (1970) 4596.
260 J. Turk and R.H. Shapiro, *Org. Mass Spectrom.*, 5 (1971) 1373.
261 N. Uccella, I. Howe and D.H. Williams, *Org. Mass Spectrom.*, 6 (1972) 229.
262 M.L. Vestal, *J. Chem. Phys.*, 41 (1964) 3997.
263 M.L. Vestal, in P. Ausloos (Ed.), *Fundamental Processes in Radiation Chemistry*, Interscience, New York, 1968, p. 59.

264   P. Vouros and K. Biemann, *Org. Mass Spectrom.*, 2 (1969) 375.
265   M.B. Wallenstein, *Ph.D. Thesis*, University of Utah, 1951.
266   A.D. Walsh, *Discuss. Faraday Soc.*, 2 (1947) 18.
267   A.D. Walsh, *J. Chem. Soc.*, (1953) 2260.
268   H. Wankenne and J. Momigny, *Int. J. Mass Spectrom. Ion Phys.*, 7 (1971) 227.
269   G.G. Wanless and G.A. Glock, Jr., *Anal. Chem.*, 39 (1967) 2.
270   W.D. Weringa, H.J.M. Sinnige and M.J. Janssen, *Org. Mass Spectrom.*, 5 (1971) 1399.
271   R. Westwood, D.H. Williams and A.N.H. Yeo, *Org. Mass Spectrom.*, 3 (1970) 1485.
272   F.A. White, *Mass Spectrometry in Science and Technology*, Wiley, New York, 1968.
273   D.H. Williams and R.G. Cooks, *Chem. Commun.*, (1968) 663.
274   D.H. Williams, R.G. Cooks, J. Ronayne and S.W. Tam, *Tetrahedron Lett.*, (1968) 1777.
275   D.H. Williams, R.S. Ward and R.G. Cooks, *J. Amer. Chem. Soc.*, 90 (1968) 966.
276   D.H. Williams, S.W. Tam and R.G. Cooks, *J. Amer. Chem. Soc.*, 90 (1968) 2150.
277   D.H. Williams, R.G. Cooks and I. Howe, *J. Amer. Chem. Soc.*, 90 (1968) 6759.
278   R. Wolovsky, *J. Amer. Chem. Soc.*, 92 (1970) 2132.
279   R.B. Woodward and R. Hoffman, *J. Amer. Chem. Soc.*, 87 (1964) 395.
280   A.N.H. Yeo, R.G. Cooks and D.H. Williams, *Chem. Commun.*, (1968) 1269.
281   A.N.H. Yeo and D.H. Williams, *J. Amer. Chem. Soc.*, 92 (1970) 3984.
282   A.N.H. Yeo and D.H. Williams, *Org. Mass Spectrom.*, 5 (1971) 135.
283   V.I. Zaretskii, N.S. Wulfson, V.G. Zaikin, V.N. Leonov and I.G. Torgov, *Tetrahedron*, 24 (1968) 2339.
284   C. Zener, *Proc. Roy. Soc. (London), Ser. A*, 137 (1933) 696.

Subject Index

Compound and Formula Index